PHYSICS WITHOUT BLINDERS
REVEALING THE ERRORS THAT CRIPPLED A SCIENCE

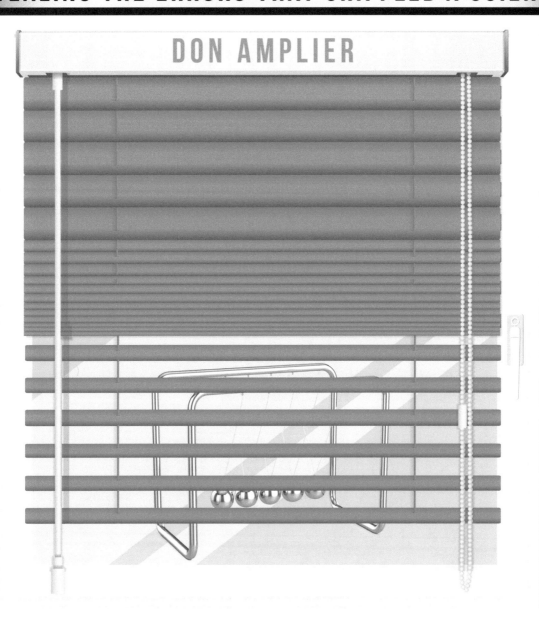

DON AMPLIER

Copyright © 2025 Don Amplier.

All rights reserved. No part of this book may be used or reproduced by any means, graphic, electronic, or mechanical, including photocopying, recording, taping or by any information storage retrieval system without the written permission of the author except in the case of brief quotations embodied in critical articles and reviews.

Archway Publishing books may be ordered through booksellers or by contacting:

Archway Publishing
1663 Liberty Drive
Bloomington, IN 47403
www.archwaypublishing.com
844-669-3957

Because of the dynamic nature of the Internet, any web addresses or links contained in this book may have changed since publication and may no longer be valid. The views expressed in this work are solely those of the author and do not necessarily reflect the views of the publisher, and the publisher hereby disclaims any responsibility for them.

Any people depicted in stock imagery provided by Getty Images are models, and such images are being used for illustrative purposes only.
Certain stock imagery © Getty Images.

ISBN: 978-1-6657-6704-0 (sc)
ISBN: 978-1-6657-6703-3 (hc)
ISBN: 978-1-6657-6702-6 (e)

Library of Congress Control Number: 2024921490

Print information available on the last page.

Archway Publishing rev. date: 2/3/2025

To A, B, C, and O

CONTENTS

Preface vii
Acknowledgements ix
Important Notes xi

Introduction xiii

Chapter 1   Newton's Laws of Motion 1
Chapter 2   Momentum's Best Friend...13
Chapter 3   Hiding in Plain Sight..27
Chapter 4   Mathematical Trickery 39
Chapter 5   Kinetic Energy's Origin Story 49
Chapter 6   The s'Gravesande Experiment 59
Chapter 7   The Cart Experiment 71
Chapter 8   The Astronaut Wrench Experiment..81
Chapter 9   Definitions 85
Chapter 10 Illusion Theory 95
Chapter 11 The Dynamic Effort Hypothesis..103
Chapter 12 The Law of Conservation of Energy 117

Appendices List 129

Appendix A Secrets to Understanding 130
Appendix B Mathematics Simplified 136
Appendix C The Scientific Method..144
Appendix D The Academic Experiment 146
Appendix E Momentum and Kinetic Energy 150
Appendix F Negative Work 152
Appendix G Historical Figures..155

Glossary 159

PREFACE

Every scientific discovery occurred because someone thought to ask the right questions. This is so true that it even applies to accidental discoveries. Take the case of Dr Alexander Fleming (1881 – 1955), who upon returning from summer vacation came across a moldy petri dish. Instead of washing it and continuing his work, he asked about the bacteria that should also have been there. After he asked a few more questions, he discovered penicillin, aka the world's first antibiotic.

In the author's case, he posed a perfectly reasonable question; he wanted to know why a particular physics formula was valid. At the time, he was re-acquainting himself with high school physics and noticed something odd. It was not that he did not understand the formula; he understood it perfectly and had no difficulty using it. The problem had to do with the premise behind the formula; it did not seem to align with reality. And to make matters worse, the book wanted its readers to take its statements on faith. It said nothing about the formula's origin or any experiment confirming it.

At first, the author's focus revolved around finding the source of the formula; where did it come from and why? Thereafter, he continued to ask other relevant questions. Some of them were answered in the books he had on hand. Others required rummaging through libraries and speaking with physicists.

The answers the author uncovered are surprising to say the least. For example, one involved noticing an experimental error no competent physicist would tolerate today. It was however an easy mistake to make in the distant past. Shockingly, that error is so egregious it is enough to foreshadow a physics revolution like no other. If any physicist thinks the author is misguided, or a fool, they need only jump to Chapter 3. There, they will encounter three scenarios, the second of which easily dispels such unflattering thoughts. And as astonishing as Chapter 3 is, it is merely a teaser for what is to come.

By now, the reader might be wondering about the book's title, "Physics Without Blinders," when questions seem to be the order of the day. The thing is that blinders have the unfortunate effect of making it difficult to see. And if you cannot see clearly, it becomes nearly impossible to ask the right questions. Anyone can ask questions but, the trick is in asking the right ones.

Blinders can also tell people there is nothing to see which, prevents them from even looking. Today, physicists run afoul of this all the time. Some fellow offers an intriguing insight into physics and summarily gets ignored. In most cases, those insights have little to no value and so, the blinder causes no harm. However, occasionally someone surprising arrives on the scene who offers a remarkable insight that changes everything. Michael Faraday (1791 – 1867) was one such individual. He was the under-educated son of a blacksmith who contributed so much to physics that Albert Einstein (1879 – 1955) kept a picture of him as a reminder.

Blinders can also prevent someone from accepting the answers to important questions. For example, Dr William Harvey (1587 – 1657) made a revolutionary medical discovery. When he revealed his findings in 1628, no one believed him. The blinders his contemporaries wore would not allow them to accept his impeccable research.

In summary, the author posed the questions no one ever thought to ask. This resulted in answers that conflict with a physicist's education. Consequently, the blinders physicists wear will be an issue. They will tell physicists to put more faith in the lessons they learned in high school than the overwhelming evidence the author provides.

ACKNOWLEDGEMENTS

No one accomplishes anything important in science without the help of others. In the author's case, he had the benefit of those historical figures described in Appendix G. Each of them, directly or indirectly, made this book both possible and necessary.

The author also wishes to acknowledge David Bodanis who wrote, "$E=MC^2$; A Biography of the World's Most Famous Equation." His book emboldened the author by corroborating his historical research.

IMPORTANT NOTES

All readers should examine Appendix A, "Secrets to Understanding," as the first step in reading this book. Public schools do not offer classes on how to learn. Teaching someone the mechanics of how to read is not the same as teaching someone how to understand. Fortunately, extremely useful techniques do exist. The most important of these can be found in Appendix A; ignore them at your own peril.

Non-physicists must avoid thinking the information presented herein is beyond them; it is not. They need only set aside any fears they might have with respect to math or physics. If someone's grasp of the simplest forms of mathematics is weak or even nonexistent, they need only read the information given in Appendix B, "Mathematics Simplified." The thing is that math is an important part of physics and so, mathematics must be included in this book. Fortunately, the author does all the calculations meaning the reader's only job is to get some small sense of the math. Believe it or not, this could be as simple as seeing which variables are in the given equations and formulas. This is an easy thing to do and it can be surprisingly informative. In short, non-physicists need not focus on the math; let physicists worry about the numbers.

Physicists should know that they will not be asked to take anything on faith but rather, they are expected to verify everything. If the standard demanded for an extraordinary claim is extraordinary proof, that standard has been met and then some.

The one thing physicists must not do is underestimate this book. Yes, it was written using simple math but, that does not mean it has nothing valuable to offer. The profound is not complicated and hard to understand; it is always simple and easy to grasp.

INTRODUCTION

Individuals who question the accepted laws of physics are generally ignored. More than that, they are often the subject of jokes and insults. On rare occasions, a kind physicist will try to correct a misguided soul. From a physicist's viewpoint, such responses are quite reasonable. They do not believe outsiders can contribute to the science of physics. So, what is someone to do if he or she did the "impossible"?

About the only thing such an individual can do is publish a book detailing irrefutable evidence and hope for the best. The other option is to submit a Paper to an official Physics Journal but, that is far less likely to succeed. Such esoteric publications never consider Papers from outsiders. Physics Journals only want Papers submitted from insiders like those working at major universities, large corporations, and government entities like NASA. Moreover, they only seem to publish Papers detailing the most obscure ideas ever recorded. The thing is that precious few of those Papers advance physics. They do however allow physicists to say they are published authors.

To encourage the physics community to examine the author's research, this book begins with a teaser. It manifests as a challenge that culminates in a remarkable revelation in Chapter 3. If it is beyond reproach, which it is, physicists will have no choice but to read on. This is of course predicated on their willingness to set aside their blinders, at least temporarily. The challenge itself goes to the validity of something surprisingly simple that every physicist believes is true.

To ensure that anyone can be a witness to a set of historic revelations, everything in this book gets explained. In simple terms, the physics only involves looking at how objects change their motion. Understanding this information will be easy, especially if the reader uses the proven techniques given in Appendix A.

The author's challenge involves something simple, well established, and still taught as fact. Specifically, it questions the validity of Newton's Three Laws of Motion. The challenge begins by reviewing each of those laws individually and then examining them as a unified package. Shortly after that, the author takes physicists somewhere they never go and reminds them of those things they seem to have forgotten. Obviously, there is no sense in re-examining Newton's Laws if we only visit the same places physicists have.

At this point, physicists are apt to be supremely confident; they know Newton's Laws have been around a long time. If those laws were wrong in any way, surely someone would have noticed long before now. On the other hand, if any physicist has ever won a bar bet, or even lost one for that matter, they might be getting a tiny bit nervous. A bar bet is where someone makes a statement that cannot possibly be true. After agreeing to a wager, the impossible happens as the bet's originator employs an unforeseeable gimmick and wins the bet.

Generally, fairness is not part of winning bar bets. For example, if someone says they can stay underwater without any breathing apparatus for ten minutes or longer, walk away. This only requires someone holding a glass of water above their head, hardly fair. As for the author's challenge, he promises no underhanded tricks or weird gimmicks; they are not needed. He does however have two rather important historical facts on his side. To begin, kinetic energy was unknown to Newton when he formulated his Three Laws of Motion. The Kinetic Theory of Heat was likewise a mystery to him. The longevity of Newton's Laws may be on the side of conventional wisdom but, the sequence in which certain key physics discoveries were made are not.

In the interest of expediency, physicists may jump to Chapter 3 if they think they know all there is to know about Newton's Laws, momentum, kinetic energy, and magic. Yes, the word "magic" was placed in the same sentence as Newton's Laws, and it is not as strange or as crazy as it sounds. Consider the one indisputable similarity between a magician and a physics teacher. Magicians have been known to say they can circumvent the laws of nature. They then demonstrate this ability, usually quite dramatically. Some physics teachers are known for performing entertaining experiments showing their students the laws of physics are indeed valid. In each case, an entertaining demonstration "proves" a claim is true.

Magicians use misdirection along with smoke and mirrors to fool their audiences. Obviously, no one believes science teachers knowingly employ such tactics; they simply pass on the accepted laws of nature. The thing is that teachers only teach what they were taught themselves. If one or more of the earliest discoveries in physics was flawed in some small way and that flaw went unnoticed way back when, they might still be teaching the same faulty lessons today. If the reader does not believe this could happen, consider the standard relationship between a student and a teacher. The student, knowing less than the teacher, has no choice but to accept anything their teachers teach. If a student knew everything the teacher knew, there would be no point in being the student.

Occasionally, a rebellious student turns up. He might question everything a political science or an economics professor says but, his skeptical attitude never follows him into the physics course room. And even if it did, physics professors are pretty sharp. They can easily debunk an uninformed student's ideas or answer any naïve questions. Moreover, these educators believe they are relaying the truth, the whole truth, and nothing but the truth. This adds up to a situation where a rebellious student does not stand a chance; a physics professor will always prevail. Most importantly, students simply do not know enough about physics to ask the right questions. And finally, there is the fact that these educators decide who passes the class and who does not. This can be a powerful incentive to tow the party line and particularly so in a physics class at an expensive university.

Physicists are aware of certain scientific missteps that went undetected for centuries. That one or two such mistakes could remain undetected today sounds impossible and yet, far stranger things happen all the time on this blue marble we call Earth. For example, the Flat Earth Society is still alive and well in the 21st century; they even have their own shiny website. We also have the odd societal situation where many athletes, movie stars, and other entertainers earn incredibly large sums of money. At the same time, many extremely valuable workers in our society sometimes struggle to make enough just to get by. And as sad as these two examples are, there are many others, some of which, are far stranger.

The thing is that if someone developed a hypothesis that appeared perfect, it might go unchallenged for quite some time. If no one thoroughly tested it due to its apparent axiomatic nature, it might become something everyone thinks is true. In short, a slightly flawed hypothesis could, under the right conditions, transform into an accepted scientific fact. Should this happen, physicists would ignore, ridicule, or try to correct anyone who dares suggests looking into its validity. It would remain "true" until the right rebellious soul comes along.

Could this have happened to Newton's Laws? Could those laws contain a flaw so small that no one noticed it before? Physicists say no and yet Chapter 3, with a small assist from Chapter 4, conclusively demonstrates the opposite is true. Before getting to that point, it is first necessary to explain Newton's Laws of Motion and one other physics concept in simple terms. As previously offered, physicists may jump to Chapter 3 if they wish. They can also review the information intended for non-physicists (Chapters 1 and 2) to make sure the author is not pulling a fast one. Who knows, the physicist who dares to examine the simplest ideas in physics explained in a simple straight forward manner might see something they had not previously considered; stranger things have happened.

CHAPTER 1

Newton's Laws of Motion

This chapter assumes the reader knows nothing of math or physics and so, everything gets explained in simple terms. If the reader runs into any difficulties, they should follow the recommendations given in Appendix A, "Secrets to Understanding." If they are uncomfortable with mathematics, they should also read Appendix B, "Mathematics Simplified." While the math given throughout this book is important, readers need not be concerned if they feel it is beyond them. For non-mathematicians, they need only read and understand the explanations the author provides along the way. These will be given in plain English. At the very least, non-mathematicians need only be aware of the variables in play. A variable is a letter or letters that represent known or unknown numerical values for such things as time, distance, and so on.

Newton's Three Laws of Motion became widely known when Sir Isaac Newton (1642 – 1726) published his book, "The Mathematical Principles of Natural Philosophy." This historic book is often referred to simply as "The Principia." It begins with a series of eight basic definitions. This is then followed by a section entitled, "Axioms, or Laws of Motion," containing the details of what is known today as Newton's Three Laws of Motion. The third edition of Newton's book was translated from Latin into English in 1729 by Andrew Motte (1696 – 1734).

During Newton's lifetime and before, European scholars were fluent in Latin. This allowed them to read existing scientific and philosophic manuscripts. It also made it possible for them to communicate with their counterparts in other nations. Today, English is the language of choice for those physicists who wish to interact with their international colleagues.

Newton's First Law

> "Every body perseveres in its state of rest, or of uniform motion in a right line, unless it is compelled to change that state by forces impressed thereon."

Using contemporary language, this law simply states that an object remains at rest (not moving) or in motion (traveling at a constant speed and in a straight line) unless acted upon by an external force.

The word "force" does not describe a mysterious or nebulous thing; it is merely the act of pushing on or pulling something. This law states that if an object changes its motion in any manner, it was pushed or pulled by an external force.

Here on Earth, proof of Newton's First Law of Motion can be surprisingly difficult to come by. Place a book on a shelf, and it will stay there until someone comes along and moves it. However, throw a book and it quickly stops moving. In space things are far different as we now know. For example, in the mid-1970s, NASA launched two space probes called Voyager 1 and Voyager 2. Since then, they have made their way beyond the limits of our solar system. They continue to move outward at a constant speed and in a straight line without the need of rocket fuel.

The biggest difference between being on Earth's surface and being in space is the atmosphere, that and the effect gravity has. As objects move about here on Earth, they push individual air molecules about causing a bit of wind. This results in objects slowing down and eventually stopping. For example, hit a golf ball and it might travel 100 yards, more if you know what you are doing. If someone were to measure the ball's speed after a golfer hits it, they would notice the ball immediately begins slowing down.

The effect gravity has is to make things fall. A golf ball might rise to a maximum height of 150 feet depending on who hit it and how hard. Thereafter, it starts dropping until it reaches the ground where it may bounce a few times before rolling to a stop. In that case, the grass acts like the air; it slows the ball down. Both the air and grass act as if they provide a passive force causing the ball to stop moving. The word "passive" indicates that the action of the air and grass in such situations can only slow the ball down, not speed it up.

All this means is that Newton's First Law is valid both in space and here on Earth. It is just easier to see objects continuing to move in space where there is no grass, air, or anything else capable of changing an object's speed. So, there is something to be said for Newton's First Law of Motion and the man who figured it out some three hundred and fifty years ago.

Newton's Second Law

> "The alteration of motion is ever proportional to the motive force impressed; and is made in the direction of the right line in which that force is impressed."

In all physics books and courses, Newton's Second Law takes the form of a simple equation. It shows the mathematical relationship a quantity of *force* has (represented by the variable "**F**") with the amount of *mass* an object has (represented by the variable "*m*") and the rate at which that object *accelerates* (represented by the variable "**a**"). A reminder, a variable is a letter or letters that represent known or unknown numerical values for such things as time, distance, and so on. The usual mathematical form for this law is **F** = m × **a**, typically written without the multiplication sign (×) as in **F** = ma.

Coming up with this equation from Newton's translated statement requires some insight. It begins with a decent understanding of the definitions Newton provides at the beginning of his book. It ends with the additional information he gives immediately after formally stating his Second Law.

Mass is a measure of an object's resistance to acceleration. A bit more crudely, mass is how much stuff in an object. The more stuff an object contains the more mass it has and thus the more it resists changing speed. It is far easier to get a toy truck moving than a full-sized one; one is measured in ounces and the other in tons.

Mass is roughly equivalent to weight. There is a bit more to know about the similarities and differences between mass and weight but for now, it is okay to assume they are roughly the same thing. This is because if an object's mass doubles, its weight also doubles. In the International System of Units (SI), the unit for mass is the kilogram (kg). Unfortunately, most consumer scales only measure weight in terms of kilograms and or pounds. Technically, neither the kilogram nor the pound is a proper measure of weight. (More about this in a later chapter.) For the record, 1 kilogram is about 2.2 pounds in the British System of Units (aka the Imperial System). This system is generally used in the United States and, to a certain degree still used in the United Kingdom.

Acceleration is the rate at which an object's velocity changes. Please note that velocity refers to how fast an object moves (speed) *plus* a notation of the direction in which it travels. If something is not moving and then begins accelerating at 1 meter per second per second (m/s^2), it means that as each second passes, the speed of that thing increases by 1 meter per second (m/s). So, after 1 second of acceleration, the object moves at 1 m/s. If the same object accelerates from rest for 2 seconds, it will be moving at 2 m/s.

For those readers who know nothing of meters, a meter is a unit of distance that is slightly longer than 3 feet. More precisely, a meter is about 39 and $^3/_8$ inches.

The basic unit for force is the newton. A single newton of force is enough to cause an object with a mass of 1 kilogram to accelerate at a rate of 1 m/s^2.

Newton's Second Law describes an important physics principle that has everyday implications. Think of an automobile; the engine provides force via the wheels and that allows a car to accelerate from a stop sign to the posted speed limit. Should someone treble the force an engine provides while keeping the car's mass the same, that vehicle would accelerate three times faster than before. If we instead reduce the car's mass to a third of what it was and use the original engine, that vehicle will still accelerate three times faster than the manufacturer intended. Automobile racers understand this; they put the most powerful motor they have in the lightest car possible. NASA and SpaceX also rely on this principle; they limit the amount of mass they send into space. Put too much stuff on top of a rocket that is not powerful enough, and it will not accelerate fast enough to make it into orbit.

Readers probably already know that everything from academic courses to zoo buildings rest on a figurative or literal foundation. Physics is no exception; it too depends upon certain basic principles and ideas. The equation **F** = m**a** is one of these and it is one of the easiest physics fundamentals to understand. In short, Newton's Second Law is merely a statement given in the form of an equation that shows the relationship *force* has with *mass* and *acceleration*. The mathematical relationship this law defines makes perfect sense; push on something twice as hard as before and that thing accelerates twice as fast.

For those uncomfortable with mathematics, the following is surprisingly simple. It should not cause any issues and even if it does, the reader can read Appendix B, "Mathematics Simplified," and then re-read the following. Should any reader not wish to engage in any math, they may jump down to the section entitled, "The Three Facts to Remember about Newton's Second Law." However, every reader should at least try to understand the following. It is far simpler than anyone might imagine.

The equation **F**=**ma** can be rearranged using algebra (the area of mathematics that deals with variables and the manipulation of equations). This results in a set of three formulas, each of which calculates the unknown value if the other two quantities are known. For example, if you know an object's mass and the amount of force that will act on it, you can predict its rate of acceleration.

The word "equation" refers to an associating of two quantities. For example, 5 = 4 + 1 is an equation that tells us that the quantity on the left-hand side of the equals sign (=) is numerically the same as the quantity on the right-hand side. Physics equations illustrate the relationship between various things and sometimes they look like a formula. A formula is also an equation but one whose purpose is to calculate an unknown quantity. The example equation (5 = 4 + 1) is not a formula since it does not seek to calculate an unknown value.

$$\boldsymbol{F} = \boldsymbol{ma}$$

Formula 1

Formula 1 calculates the amount of force acting on an object if we know how much mass it has and how fast it accelerates. In this case, the variable "**F**" represents the unknown quantity.

$$\boldsymbol{m} = \frac{\boldsymbol{F}}{\boldsymbol{a}}$$

Formula 2

Formula 2 calculates the amount of mass an object has when we know the amount of force acting on it and can measure its rate of acceleration.

$$\boldsymbol{a} = \frac{\boldsymbol{F}}{\boldsymbol{m}}$$

Formula 3

Formula 3 predicts the rate at which an object will accelerate if we know the amount of mass it has and the quantity of force that will act on it.

Please note that the variables for force and acceleration (**_F_** and **_a_**) are in **_bold italics_**. The variable for mass (_m_) only uses _italics_, letters that slant to the right. The use of italics lets you know that the letters used in mathematical equations and formulas represent variables. A variable is a letter that represents numerical amounts or quantities for such things as force, mass, acceleration, and other things. Stated differently, variables are mathematical stand-ins and are used for both known and unknown values. In many cases, the first letter for something is the ideal means of writing it mathematically. For example, the variable for "mass" comes from its first letter "m." In this book, **bolding** a variable tells you that the variable also contains a directional component and is therefore a vector. Please note that there are other methods of showing a variable is a vector, none of which are used in this book. A vector is something that has both a quantity (aka a magnitude, how much of something exists) and a notation of direction attached to it. If something only has a quantity, it is scalar. For example, mass is scalar since it never has direction associated with it. When the reader sees the word "scalar," think of the word "scale." Scales measure weight; they produce values with no hints or thoughts of direction.

Force and acceleration could also technically be pure quantities (scalars) having no notation or mention of direction. Unfortunately, physicists demand the variables for force and acceleration have some notation for direction. This tends to complicate physics since vector mathematics has its own sets of rules. These will be explained in simple terms when necessary. Fortunately for non-physicists, the equations and formulas in this book are of the simplest type and thus the rules that apply to vectors are easy to grasp. This means vectors will not be an issue the reader need fear.

The only thing left to know about any set of physics formulas or equations is that the units of measure must always be in accord. As stated previously, the basic unit for force is the newton. And for this unit and the three formulas to be valid exactly as written, acceleration must be in meters per second per second (m/s^2) and mass must be in kilograms (kg).

To make this law a bit more relatable, take the case of a four-kilogram object at rest. That object weighs nearly the same as a ten-pound bag of potatoes. In this example, complicating issues such as gravity, air, or grass do not exist. If we apply an eight-newton force acting in the northerly direction and apply that for exactly one second, what will happen?

To begin, our four-kilogram object will accelerate in the same northerly direction. This is because objects always move in the same direction they get pushed. We can calculate how fast the object accelerates using Formula 3.

$$a = \frac{F}{m}$$

Formula 3

Substituting the known values for force (**F** = 8 newtons) and mass (*m* = 4 kilograms) gives us the following.

$$a = \frac{8}{4}$$

$$a = 2 \text{ meters per second per second}$$

Calculating Acceleration

Because force only acts for 1 second, it becomes extremely easy to calculate the resulting velocity. It will be traveling in a northerly direction at 2 meters per second.

As for memorizing Newton's Second Law or any of its mathematical forms, there is no need; there will be no test, no exam. The reader may refer here any time he or she wishes.

The Three Facts to Remember about Newton's Second Law

At the very least, readers should understand the following three points. Without using any math, they convey an elementary understanding of an important physics principle.

1. *Force* is a push or a pull that tries to or causes acceleration. You can push on a car with the parking brake set and it will not move. Push on the same car with the brakes released and it will begin to move; it will begin moving faster and faster.

2. *Mass* is a measure of an object's resistance to acceleration. The more mass an object has, the slower it will accelerate for a given amount of force. If you were to take the same amount of force to get a full-sized sedan moving and apply that to a toy car instead, that toy would accelerate far, far more quickly.
3. *Acceleration* is the rate at which objects change how fast they move. Things can accelerate positively and thus speed up. They can also accelerate negatively (aka decelerate) and thus slow down. Stated differently, acceleration occurs whenever objects change their velocity. Velocity is the combination of speed and the direction of travel.

Newton's Third Law

"To every action there is always opposed an equal reaction: or the mutual actions of two bodies upon each other are always equal, and directed to contrary parts."

This law is often paraphrased as "for every action there is an equal and opposite reaction." It is further explained by physics professors in terms of force; they say force always acts in two directions simultaneously. This law explains how and why rockets work. Basically, a rocket pushes gas out of its tail while at the same time an equal force accelerates the rocket in the opposite direction.

Please note that rockets burn fuel which releases the chemical energy necessary to produce force. That burnt fuel then becomes propellant, the gas forced out the back of a rocket so that the rocket itself accelerates in the opposite direction.

Because force always acts on two different objects but in opposite directions, the opposing forces always act for the same amount of time.

A funny little story about this law and rockets involves a newspaper article written about a hundred years ago. A reporter who did not understand Newton's Third Law of Motion mocked a famous American rocket pioneer. The reporter wrote that rockets cannot function in the vacuum of space. He incorrectly stated that rockets work by pushing on the atmosphere. Sometimes an authority figure (the newspaper man in that case) does not know as much as they think they do.

The Two Facts to Remember about Newton's Third Law

1. When force acts on one object, an equal force acts on a different object but in the opposite direction.
2. The time force acts on one object is exactly equal to the time force acts on the other.

Momentum

Momentum is a numerical value equal to the product of mass (*m*) and velocity (**v**) written mathematically as *m* × **v** or, more usually, *m***v**. Velocity is how fast something moves in relation to something else and in what direction. The letter "v" typically represents velocity and because it contains a directional factor, it is a vector and its variable is written in bold type (**v**). Mathematically any vector, like velocity in this case, multiplied by a scalar like mass produces another vector moving in the same direction. Momentum is therefore also a vector.

The letter "p" typically denotes momentum. Why not "m" the reader might ask? The answer beyond tradition or that "p" comes from some Latin word is that the letter "m" has other uses, two of which are already part of momentum. The first use relates to mass which, is part of the mathematical form for momentum (*m* × **v**). The letter "m" also references the word "meters" in meters per second and is the preferred scientific means of quantifying speed or velocity. Please note that the letter "m" is only italicized when used as the variable for mass.

As a brief aside, the need for the letter "p" to describe momentum may be unnecessary. The formula for momentum consists of just two variables, "*m*" and "**v**"; it is just as easy to use the letter combination "mv" to refer to momentum. If this were done, it would free up the letter "p" for some other use. This is not a proposal, just a random thought the author felt he could share. Moreover, it is not like there are no existing variables in physics that use two letters. Humorously, if the author loses the bar bet mentioned in the Introduction, he claims credit for using the letter combination "mv" to describe momentum instead of the letter "p".

The usual formula for momentum is as follows.

$$\boldsymbol{p} = m\boldsymbol{v}$$

Formula 4

Formula 4 says the amount of momentum (*p*) an object has is equal to the amount of mass (*m*) it has multiplied by its velocity (*v*).

The Scientific Method and Newton's Laws of Motion

The scientific method, described in Appendix C, constitutes a significant part of this book and so, it will be mentioned from time to time. It is supposed to be the method for determining the worth of new and existing scientific ideas. Every new hypothesis (a proposed description of a phenomenon of nature) must have a testable prediction. If it has none, its validity becomes a matter of faith. This is because if there is nothing to test, there is no way to know if the hypothesis is valid. Only when a hypothesis makes a prediction, and it is found true by enough scientists can it become an accepted theory (the basic information considered true that describes the inner workings of a phenomenon of nature).

The scientific method has not always been standard practice. A long time ago, if a scholar proposed an idea and it seemed to make sense, others would often accept the proposal. And even if other scientists thought to test something experimentally way back when, they did not necessarily have the means to measure anything to any degree of precision. The margin of error in measuring velocity, time, and other things was so great in the distant past no one could be certain of anything, not really.

Newton's Three Laws of Motion, when combined, make a prediction thereby satisfying the first requirement of the scientific method. They predict another law and one that can be tested relatively easily today. That law is called the Law of Conservation of Momentum. It states that in any closed system, zone, or area, the amount of momentum (mathematical quantity of motion) inside that zone will remain unchanged regardless of what occurs therein. The only proviso is that nothing from without interferes. In other words, if a known mathematical quantity of motion (momentum) exists within some zone or area and then some action occurs, the original quantity of motion will continue to exist. This is called conservation because the amount of something remains unchanged. For example, if we have two objects, one stationary and the other moving toward it, the amount of motion before they collide will have a certain mathematical value. After the collision occurs, the amount of motion remains mathematically the same which, means momentum is conserved. Likewise, if some closed system of objects has no momentum, any action that later occurs within that area will not change the amount of momentum.

Some teachers skip over how Newton's Three Laws of Motion combine to produce the Law of Conservation of Momentum; they just tell students those laws combine in that way. Other teachers let students in on the mechanics of how Newton's Laws work together. It begins by pointing out that Newton's First Law states that the quantity of momentum an object has will remain unchanged until acted upon by an external force. Because a moving object has mass and moves at an unchanging velocity, it must retain the amount of momentum (mass × velocity) it has until such time as force acts. Newton's Second and Third Law maintain that force which, is necessary to alter velocity and thus momentum, acts predictably and uniformly on both objects. The predictably aspect ties to Newton's Second Law (***F=ma***). The opposite directions and uniform factors relate to Newton's Third Law; "for every action there is an equal and opposite reaction." So, if one object collides with another, the amount of momentum one object receives is equal to the amount of momentum the other loses and of course, vice versa. In this way, the overall momentum is then said to remain unchanged. The amount of momentum exchanged during collisions depends on how much mass each object has and their individual velocities.

If anyone could show an example where momentum is not conserved, it would mean that Newton's Laws of Motion make a prediction that is not true. This would be enough to invalidate one or more of those Laws and would mean physicists were improperly taught by others who had the same misfortune. In such an event, the very foundation of physics as she is taught today would crumble. Textbooks would have to be tossed out and replaced with those yet to be written. Physics courses would come to a screeching halt while physics professors scrambled to find something they could still teach. And there would be other consequences as well, the details of which are too gruesome or wonderful to list here. In short, showing a flaw exists within Newton's Laws of Motion would be a surprisingly big deal; it would be an extremely rare and historic event.

The last thing to know about momentum is that it has no named unit despite being part of physics for more than 350 years. Mass has a unit, the kilogram. Distance and time have their own units, meters and seconds. The unit for momentum is just the units that make it up, kilogram meters per second. This is very odd since momentum's best friend has its own named unit. Interestingly, this friend became part of physics after momentum's value had already been accepted by all.

By now the reader should have the idea that Newton's Laws are infallible and that they make perfect sense. This suggests the author should have chosen something far easier to challenge. The thing is that there are still certain key bits of information to pass on that physicists always undervalue or neglect in certain common situations. For non-physicists, they must first learn just a little bit more, this time about momentum's best friend.

CHAPTER 2

Momentum's Best Friend

Decades after scholars recognized momentum, a related mathematical expression began rising in importance. Over time, it and momentum became best friends, intimate companions as it were. They share several things in common the first of which revolves around the variables they use; they are the exact same ones. Moreover, both of their formulas can be derived from Newton's Second Law of Motion. Yet another commonality is that physics students always learn about both expressions before studying collisions.

Collisions

Collisions were among the first things studied by 17th century scientists and so, they carry a certain historical importance. Surprisingly, collisions are still a significant part of the subject even today. Some readers may have heard of the Large Hadron Collider (LHC) going on-line more than a decade ago. At that time, some individuals feared it might create a black hole; it did not. The LHC is the most expensive and largest machine in the world. Ironically, it investigates collisions that occur with the smallest objects known to man, i.e., atoms and their constituent parts, aka hadrons.

A collision occurs when two or more objects move towards one another and interact, often with a measure of violence. They come in only two types and yet, some real-world collisions contain elements of both. The first type is called an *elastic* collision. For example, a baseball bat striking a baseball is an elastic collision between the bat and the ball. It is called an elastic collision because before they meet, each object moves in its own unique way. After the collision is over, each still moves in its own unique way albeit differently than before. In other words, a bouncing or elastic effect occurs between the colliding objects. Other examples include such things as dribbling a basketball, a child kicking a clown, and anything similar including one boxer punching another. In all cases, two or more objects move towards one another, collide, and then each moves differently than before.

The second type of collision starts out the same in that two or more objects move towards one another. As they meet, the objects unite in some fashion and then move off as a single object; they now travel at the same velocity in other words. Such collisions are called *inelastic* because no bouncing effect occurs. For example, an inelastic collision occurs whenever anyone catches a baseball in a glove. This begins with two objects moving in their own unique way and ends with a single object; a ball inside a glove where the ball moves in the same way the glove does. Other examples of inelastic collisions include such things as a dart thrown into a target, a linebacker in football tackling a running back, and anything similar including when extraterrestrial objects collide with the Earth. In all cases, two or more objects collide to form a single larger one.

One way to think of an inelastic collision is to think of it as being an explosion that unfolds in reverse. For example, a loaded rifle begins as a single object that includes a projectile and a quantity of gunpowder. Before pulling the trigger, the rifle and everything within it move as if they were a single object. After a controlled explosion is over, the projectile flies off in one direction while the rifle itself recoils on the shooter. One object transforms into two or more objects, each going their separate ways. This is the exact opposite of an inelastic collision where two or more objects unite in some manner to form a single one.

The reverse of an elastic collision is just another elastic collision. So, if someone videos two objects experiencing an elastic collision, someone else watching it in reverse still sees an elastic collision.

Energy

Energy is an idea with which everyone should have some familiarity. The United States Federal Government has an entire department devoted to the concept. The US Department of Energy tends to focus on fossil fuels, alternative energy sources, and other related issues. Consumer wise, we have such things as energy drinks. We also tend to care about the price of gasoline, the source of energy most automobiles require. Energy is everywhere and for good reason. Nothing in this universe can change in any way unless energy plays a role. Moreover, energy itself changes forms when used; it can also be stored in many ways.

Energy has various definitions, one of which is related to how physicists quantify it in one of its many forms. A far better definition might be to say that *energy* is basically that which can create change; it is essentially the concept we call *cause*. Whenever an object moves, it first took energy to change its status from an object at rest to something that is in motion. Once the object is in motion, it has energy because it moves; it is continually changing its position in space. The faster it moves, the greater the changes in distance per second and thus the more energy it will have.

Kinetic Energy, typically abbreviated "*ke*", is an example of an existing physics variable that uses more than one letter. The word "kinetic" refers to the energy of moving objects (aka mechanical energy). The kinetic energy formula, listed below as Formula 5, illustrates that it does in fact use the exact same variables as momentum (*m* and *v*). The important difference is that one of the variables is treated differently. Velocity (*v*) has the number "2" above and just to the right of it. This means velocity gets squared (multiplied by itself). If the numerical value of *v* were 5, *v*² would be *v* × *v*, 5 × 5, or 25.

$$ke = \frac{m \times v^2}{2}$$

Formula 5

Which is the same as

$$ke = \frac{m \times v \times v}{2}$$

Typically written as

$$ke = \frac{mv^2}{2}$$

or

$ke = ½mv^2$

when it appears within the text.

Formula 5 provides a numerical value for the energy an object has because it moves. It does this by multiplying the object's mass by the square of its velocity (**v** × **v**) and then dividing the result by 2. Please note that due to the rules of math, the same answer results if the mass gets divided by 2 and that value gets multiplied by the square of its velocity (**v** × **v**).

The mathematical result of the kinetic energy formula is not a vector; it is scalar. This is despite the formula having a couple of vectors (**v** × **v**) in its formula. One of the rules when dealing with vectors is that when a vector is squared (multiplied by itself as is the case with velocity), direction vanishes from the result. This produces a pure quantity, aka a scalar.

Momentum and Kinetic Energy

When an object accelerates, force is the cause. As an object increases its velocity, time passes, and the object moves through a displacement. *Time* is scalar and uses the letter "t" as its variable. *Displacement* is distance with the notation of direction; it is therefore a vector and its variable is in bold type (**d**).

Think of an automobile accelerating after the traffic light turns green. It takes time to increase its speed; no vehicle can go from standing still to 45 miles per hour (mph) instantly. As the automobile increases its speed but before reaching the posted speed limit, it travels through the intersection and down the road a bit; it travels a certain distance. In short, whenever anything accelerates, or decelerates for that matter, that action occurs over a distance and time passes.

Momentum relates directly to the time (*t*) force acts, and this is called impulse. Mathematically, *impulse* is force × time, **F** × t, or **F**t. Impulse is a vector because force is a vector that gets multiplied by a scalar; time is scalar.

Physicists sometimes define energy as the ability to do work. Mathematically, *work* is force × displacement, **F** × **d**, or **Fd**. Work contains two vectors (**F** for force and **d** for displacement) but, the result of multiplying these two vectors is a scalar. Please note that for mathematical and other reasons, physicists use the word "displacement" when being technically precise about work. Informally, they say "work is force × distance." Both words, "distance" and "displacement" will be used in this book somewhat interchangeably depending only on the author's mood. If physicists can get a little sloppy, so can the author. Physics is what it is; a little messed up as the reader will soon come to learn.

Assuming an object begins at rest, the following two equations describe the relationship of force with the formulas for momentum and kinetic energy.

$$\boldsymbol{F}t = m\boldsymbol{v}$$

Equation 6

In words, Equation 6 says that the amount of impulse (force × time) applied to an object that begins at rest equals the momentum it will get.

$$\boldsymbol{F}\boldsymbol{d} = \frac{m\boldsymbol{v}^2}{2}$$

Equation 7A

In words, Equation 7A says the work (force × displacement) done to an object that begins at rest equals the kinetic energy it will get.

Because force and displacement are vectors whose directions are the same, direction vanishes from the result when they are multiplied together. This means Equation 7A remains balanced with respect to vectors; the left- and right-hand sides of Equation 7A result in scalars, pure quantities. Interestingly, Equation 7B gives the exact same answers as Equation 7A. Equation 7B uses no vectors; "F" is just the amount of force, "d" is the distance force acts through (no direction included and consequently no bolding of the variable), "s" is speed, and "m" is what it always is (mass). Despite this, the physics community insists on using vectors in the formulas for work and kinetic energy.

$$Fd = \frac{ms^2}{2}$$

Equation 7B

This is the point where the physics community misses the first clue something is horribly wrong with physics as she is taught. They describe a moving object in two different ways using two different mathematical expressions ($m\mathbf{v}$ and $½m\mathbf{v}^2$) that use the exact same variables.

Mathematically, momentum increases at the same rate velocity changes. With kinetic energy, it is an entirely different story because velocity gets treated differently; it gets squared ($\mathbf{v}^2 = \mathbf{v} \times \mathbf{v}$). This means when the velocity of an object doubles (× 2), its momentum also doubles but its kinetic energy increases by a factor of four (2 × 2). If an object's velocity triples (× 3), its momentum also increases by a factor of three. With a tripling of velocity, kinetic energy increases by a factor of nine (3 × 3).

No physicist can clearly articulate the conceptual differences between momentum and kinetic energy; they can only differentiate between them using mathematics. Oddly, physicists do not have a problem with this; they are quite happy using mathematics only. They claim mathematics is the language of physics. The thing is that if you can only describe a physics principle using mathematics, you probably do not understand it, or it is wrong.

If asked to list the differences between momentum and kinetic energy, a physicist always gives mathematical related responses. For example, a physicist might state that momentum increases directly with velocity while kinetic energy does not. He might point out that one is directional, and the other is not. The best they can do is to say momentum is always conserved but kinetic energy is not. However, energy in general is supposed to be conserved. This means that this so-called explanation is not helpful. A particularly clever physicist might point out that kinetic energy can transform into other forms and then say momentum cannot. The thing is that if you remove kinetic energy from something, its momentum also goes away. This is because both use the same variables (mass and velocity); so, even this evasive explanation lacks any real value.

If physics as she is taught is indeed valid, there should be a simple non-mathematical way to describe the differences between these two concepts. Ask a ten-year-old and he or she can tell you the difference between a police officer and a fireman. These two professions are also intimate companions; both work for public safety, require special training, and have their own unique equipment. Moreover, they work together and separately which, physicists know also describes momentum and kinetic energy. Humorously, members of one group pose for calendars and the other group is known for something else entirely; they hang out in donut shops.

Obtaining the impulse-momentum equation (**Ft** = m**v**) from Newton's Second Law is relatively simple. It only requires factoring (multiplying) time into both sides of the equation **F** = m**a** and knowing the mathematical relationship of time and acceleration with velocity.

To obtain the work-kinetic energy equation (**Fd** = ½m**v**^{2}) from Newton's Second Law is slightly more involved. It is not something the reader need see now, or ever. It amounts to manipulating Newton's Second Law by adding the concept of displacement (distance with the notation of direction). This often includes adding explanations describing the why behind the math. Interestingly, it usually begins with the impulse-momentum equation (**Ft** = m**v**) and moving forward from there. This is how the famous Scottish physicist and mathematician James Clerk Maxwell (1831 – 1879) showed readers how to derive the kinetic energy formula in his 1877 book, "Matter and Motion."

Special Notes

The mathematics given throughout this book has value and so, it must be provided. Non-mathematicians need not be concerned if the math seems hard to follow. If they understand the explanations given along the way, they will be okay. As for the equations and formulas themselves, just note which variables they use. This should impart a good sense of the math and what is happening. The math and the calculated values are correct. If non-mathematicians do not trust the author, have a knowledgeable friend verify them.

The author assumes physicists will verify the math and other statements; in fact, this is the second time he insists they do so. There is no intent to deceive; if any errors escaped the author's detection, they are minor at best. And even if a few snuck by him, they will not be enough to invalidate any of his conclusions.

Collision Theory

Please note that collisions can begin in different ways. Both objects could be moving in opposite directions as shown below, both could be moving in the same direction with one moving fast enough to catch up to the other, or one of the objects might not be moving at all. Collisions can also occur with objects moving at different angles to one another. For example, take the real world case of two cars colliding at an intersection; one might be traveling east and the other north. In that case, the cars are traveling at an angle of 90 degrees to one another. Such collisions are a bit more involved and will not be discusssed in this book. They only complicate matters without adding anything of value to the book's purpose.

In physics, the idea of a closed system or an isolated system is occasionally mentioned in concert with the idea of conservation. This is just a designated zone or defined area in which nothing enters or leaves. Should some physical attribute such as energy not change regardless of what happens within that zone, that attribute has the status of being conserved.

The following graphic shows two objects that are about to experience an elastic collision. This isolated system begins with only these two objects and a known quantity of momentum.

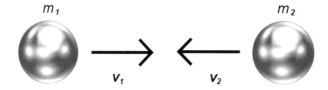

Before an Elastic Collision

Object m_1 begins with a known amount of momentum ($m_1\boldsymbol{v}_1$). The number "1" attached to the variables for mass and velocity is called a *subscript*. From the graphic above, it should be obvious that the subscript "1" refers to the ball on the left. Subscripts are always just to the right and slightly below the variable to which they belong. They are often numbers but, they can also be a letter, descriptive words, or abbreviations. This allows physicists to use the same variables (like m, $\boldsymbol{a}$, $\boldsymbol{F}$, and so on) numerous times, each with a different identifying subscript. And by always using the same variable, like "m" for mass, it becomes easier for physicists and others to understand equations, think a helpful consistency. In the case of mass, the variables m_1 and m_2 not only refer to the amount of mass an object has but they can also be used to refer to the object itself.

The second object designated with the subscript "2," has its own quantity of momentum ($m_2\boldsymbol{v}_2$).

Once this elastic collision concludes, m_1 will be moving away from the point in space where the collision happened. The same will be true for the second object (m_2).

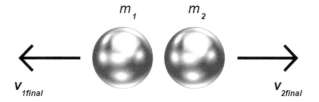

After an Elastic Collision

Please note that to distinguish between the same concept before and after an event, typically velocity, physicists use a different letter or employ some other means. This distinction is particularly important when examining collisions because the velocities of colliding objects always change. In this book, the author uses only one of the available methods; specifically, he uses a subscript. In some cases, subscripts are defined in the accompanying text. In others, a subscript's meaning is so obvious that no description is necessary.

Newton's Three Laws state that momentum is always conserved. Equation 8 shows how to write this mathematically for elastic collisions. The left-hand side of the equals sign indicates the amount of momentum that exists before the collision begins. The right-hand side describes the quantity that remains afterwards. In this equation, v_1 is the velocity of the object m_1 before it collides; v_{1final} is the velocity of that same object after the collision takes place. Similarly, m_2 begins with an initial velocity of v_2; it ends with a final velocity of v_{2final}.

$$m_1 v_1 + m_2 v_2 = m_1 v_{1final} + m_2 v_{2final}$$

Equation 8

Please note that when seeing longer equations such as Equation 8, the reader has nothing to fear. Such equations simply contain several individual expressions ($m_1 v_1$ for example) that have already been explained. Equation 8 shows the momentum ($m_1 v_1$) of object m_1 added (+) to the momentum ($m_2 v_2$) of the second object (m_2) on the left-hand side of the equal sign. The same is true on the right-hand side except this time the variables v_{1final} and v_{2final} describe the post collision velocities. Mass does not change and so, we can reuse the original variables on the right-hand side (m_1 and m_2).

Science teachers often show their students Equation 9. It states the kinetic energy (½mv^2) of object m_1 added to the kinetic energy of m_2 before the collision occurs is equal to the amount that exists within those objects afterwards. This equation assumes the objects involved are perfect or ideal; no kinetic energy is said to transform into heat or any other form during the collision. Please note that the variables for velocity have *both* subscripts and exponents. As before, the exponent *"2"* indicates velocity is multiplied by velocity and the subscripts reference a particular object.

$$\frac{m_1 v_1^2}{2} + \frac{m_2 v_2^2}{2} = \frac{m_1 v_{1final_1}^2}{2} + \frac{m_2 v_{2final}^2}{2}$$

Equation 9

Equation 10 represents the Law of Conservation of Energy for Elastic Collisions. It indicates energy can and does change from kinetic into other forms as represented by the variables E_1 and E_2. Please note that the letter "E" is an obvious abbreviation for energy. Equation 10 applies to real objects; not mythical ones as is the case with Equation 9.

$$\frac{m_1 v_1^2}{2} + \frac{m_2 v_2^2}{2} = \frac{m_1 v_{1final}^2}{2} + \frac{m_2 v_{2final}^2}{2} + E_1 + E_2$$

Equation 10

The Law of Conservation of Energy states that energy cannot be created or destroyed. The amount that exists in any closed or isolated system can never increase or decrease. Physicists extend this to mean that no energy in the universe can ever be destroyed and the creation of new energy is not possible.

Mathematically, the only difference between Equation 9 and 10 are the E_1 and E_2 variables included with Equation 10. Their presence means that some kinetic energy changes into another form during an elastic collision. In the simplest case, the E_1 and E_2 variables represent the thermal energy (heat) increase that occurs within the colliding objects. If other manifestations of energy occur, such as a flash of light or a sound, we could simply add more *E* variables or just include those quantities within the E_1 and E_2 amounts.

When Equation 8 and 9 are combined in just the right way, they produce two formulas. These calculate the resulting velocities for objects experiencing an elastic collision; one is for v_{1final} and the other is for

v_{2final}. Some educators have their students derive these formulas themselves; others provide them. These are then used to do physics problems related to elastic collisions.

To non-mathematicians, combining Equations 8 and 9 to produce the formulas for v_{1final} and v_{2final} probably appears rather intimidating. Readers need not be concerned; those formulas are given in the next chapter. Spoiler alert, they will be used to expose a lie science teachers do not know they are passing on to the next generation of physicists.

By now, the reader should see why so many students avoid physics; it gets scarier the more you learn about it. Despite this, do not allow any of the math to intimidate you. It is only necessary to understand what variables are in play in the given equations.

For the record, it is not possible to combine Equations 8 and 10. This is due entirely to the extra two variables E_1 and E_2 within Equation 10. If it were possible to produce a set of formulas using Equations 8 and 10, they would not work for every type of material. A formula that works for steel balls would not necessarily work for balls made of some other material such as ivory, glass, rubber, and so on.

Equation 11 shows the mathematical form for the Law of Conservation of Momentum for inelastic collisions. The left-hand side shows the momentum of m_1 added to the momentum of m_2. This represents the total momentum of the two objects before the collision occurs. Because the result of an inelastic collision is the combining of two objects into one, the right-hand side shows the mass of m_1 added to the mass of m_2. Thereafter, that value gets multiplied by the final velocity (v_{final}) of both objects moving as one.

$$m_1 v_1 + m_2 v_2 = (m_1 + m_2) \times v_{final}$$

Equation 11

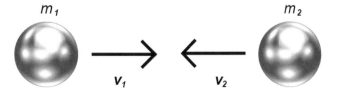

Before an Inelastic Collision

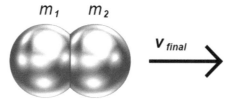

After an Inelastic Collision

Typically, the final velocity of the combined objects is unknown. To calculate it, students need only rearrange Equation 11 to produce the following formula. This is usually a snap for any student who has mastered the simplest aspects of algebra.

$$v_{final} = \frac{m_1 v_1 + m_2 v_2}{m_1 + m_2}$$

Formula 12

The amount of kinetic energy before an inelastic collision can be quite staggering. However, the quantity that exists afterwards is always less; it can even be zero. It all depends on the mass of the individual objects, how fast each moves initially, and how they combine when the collision occurs. Mathematically, kinetic energy is never used to solve inelastic collision problems; it is not needed and is of no predictive value.

Types of Kinetic Energy

Kinetic energy exists in three different but related forms. Each of these involves an object having mass that moves in some manner. The first of these is as implied earlier and involves objects moving in a straight line through space over time. Examples abound; they include such things as an automobile traveling down a road, a golf ball rolling towards a hole, and anything even remotely similar. For example, air has mass and when it moves, we call it wind. Therefore, wind is an example of kinetic energy at work (pun intended).

As for the second type of kinetic energy, it exists within all objects to one degree or another. Physicists refer to it as the "kinetic theory of heat" which, describes the mechanical aspects of thermal energy. Physically, it manifests as individual vibrating atoms and molecules. In short, thermal energy is nothing more than random molecular motion. The faster molecules vibrate, the more kinetic energy each one has, and thus the hotter the object is. Within all solid objects, molecules can only vibrate in place. This is because each molecule is trapped by the others surrounding it. Within liquids and gases, molecules are freer to migrate and commonly do but, they also still vibrate.

Sound is similar to thermal energy in that it is the result of vibrating objects such as a plucked guitar string. The difference is that sound travels through a medium such as air.

An apparent paradox occurs when touching two different objects that are at the same temperature. Consider touching both a metal pipe and a rubber strap. If both are at the same freezing temperature, the pipe feels far colder than the rubber. This is because metals conduct heat easier than rubber. And the heat in this situation is the heat within a person's hands. One feels the cold because the thermal energy in one's hand is leaving and moving into the metal pipe. In short, not all objects conduct or accept thermal energy to the same degree. This is a good thing; think gloves, coats, hats, and so on.

The third type of kinetic energy involves rotation. This includes such things as an orbiting satellite, a rotating planet, and a spinning wheel. These and other related examples have kinetic energy that is rotational in nature.

Momentum and the Three Types of Kinetic Energy

Any object that has mass and is in motion has a measure of both kinetic energy and momentum. There is no such thing as a moving, vibrating, or rotating object of any kind or size that only has kinetic energy, or only has momentum. If an object has one, it must also have a measure of the other.

The numerical difference between kinetic energy and momentum is always the same. It is a factor equal to the velocity of the object divided by two ($v/_2$). The following equation shows how the momentum of any individual object relates to its kinetic energy. This is the mathematical reason why an object that has kinetic energy must also have a measure of momentum and of course, vice versa.

$$m\boldsymbol{v} \times \left(\frac{\boldsymbol{v}}{2}\right) = \frac{m\boldsymbol{v}^2}{2}$$

Equation Relating Momentum with Kinetic Energy

In words, if you multiply the amount of momentum an object has by its velocity and divide by 2, the result is the numerical value of the object's kinetic energy.

At the molecular level, the $^1/_2$ factor still applies but only to the individual vibrating molecules. Multiply a molecule's momentum by its $^1/_2$ factor and it provides the numerical value for its kinetic energy. Add up the kinetic energy of all the vibrating molecules in any object and it produces the numerical value indicating how much thermal energy the object has.

Add up the momentum for all the individual vibrating molecules *within* an object and it produces a numerical value of zero, no net or overall momentum. This is due entirely to the directional nature of momentum. The theory for this is explained in Chapter 4. There, the reader will have no difficulty understanding the directional nature of momentum and why it can add up to a zero value.

Rotating objects also have momentum which, physicists call angular momentum. Some physics teachers like to do experiments that demonstrate angular momentum as a means of engaging students. This is because such experiments contain a "wow" factor that usually impresses young minds.

Summary of Momentum, Kinetic Energy, and Collisions

1. Collisions are either elastic or inelastic. Despite this, some real-world collisions contain elements of both. For example, if two automobiles were to collide just so, their mirrors might fly off in different directions while the two cars are fused together in some manner.
2. An explosion is tantamount to an inelastic collision unfolding in reverse and, vice versa.
3. Any object that moves in any manner has a measure of both kinetic energy and momentum.
4. Momentum and kinetic energy use the same variables in their formulas.
5. Under the right conditions, the amount of momentum two objects have, including extremely small objects such as atoms and molecules, can add up to a numerical value of zero, no net or overall momentum. For example, this can occur when two identical objects travel at the same speed but in opposite directions.

CHAPTER 3

Hiding in Plain Sight

This chapter illustrates, both analytically and experimentally, why momentum is not a conserved quantity. This should not be necessary given the fact physics professors already teach students that momentum is not conserved. This occurs during one of the more advanced lessons. At the risk of short handing that lesson too much, the Laws of Thermodynamics describe a concept called *entropy*. It describes how much disorder exists within a system. And with every event, such as an elastic collision, the amount of entropy increases. In other words, there are no perfect interactions and yet, Newton's Three Laws of Motion predict perfection does exist. This is an obvious conflict students never recognize due to how poorly physics instructors were taught to teach the subject. They begin with Newton's Laws of Motion, ideal objects, and elastic collisions. In later semesters, they cover the Laws of Thermodynamics that deal with real objects. No one does the math that adds Newton's Laws to the Laws of Thermodynamics.

When teaching Newton's Laws, some high school science teachers use a Newton's Cradle to demonstrate conservation of momentum. Typically, the teacher begins by pulling back a single ball and releasing it. This results in a single ball popping out the far side. When the teacher pulls back and releases two balls, two balls exit. Should the teacher pull back three balls, three balls leave. Teachers see this as an effective demonstration; the ball for ball exchange always seems to impress students. Sadly, everyone ignores the fact a Newton's Cradle makes a clicking sound. This begs a surprisingly important question; does that clicking noise affect conservation of momentum in any way?

Newton's Cradle

The answer begins by looking into the theory behind the ball for ball exchange. This initial scenario is unlikely to impress most physicists but, all journeys of discovery must begin somewhere. The second scenario analyzes an everyday occurrence that happens more times than anyone can count. It explains why momentum cannot be a conserved quantity. This scenario also includes the details of a revealing experiment physicists might wish to perform, especially if they have access to a laser. The last scenario is a simple but ingenious experiment designed for use in a high school physics lab.

Scenario 1, An Elastic Collision

On the surface, an elastic collision is a relatively simple event. It begins when two objects move toward one another, exchange momentum and kinetic energy, and then go their separate ways. When we look deeper, things begin to get a bit more involved.

Before an Elastic Collision

This isolated system containing only two objects begins with a known quantity of momentum ($m_1 v_1$). Object m_2 has none since it is at rest, not moving. Kinetic energy wise, only m_1 has any since it is the only thing in motion.

By manipulating the conservation of momentum and kinetic energy equations from the previous chapter, two new formulas emerge. They predict the final velocities (v_{1final} and v_{2final}) for the objects involved in an elastic collision. These formulas also tell us that the final velocities for the colliding objects depends on their initial velocities, the amount of mass each has, and nothing else. This is because these formulas only contain the variables for mass and velocity.

Important Note

Readers need only understand that the following two formulas exist and be aware of the only two variables they use (m for mass and **v** for velocity).

$$v_{1final} = \left(\frac{m_1 - m_2}{m_1 + m_2}\right) v_1 + \left(\frac{2 m_2}{m_1 + m_2}\right) v_2$$

Formula 13

$$v_{2final} = \left(\frac{m_2 - m_1}{m_1 + m_2}\right) v_2 + \left(\frac{2 m_1}{m_1 + m_2}\right) v_1$$

Formula 14

Consider the situation where the mass of m_1 is the same as m_2 while m_1 moves at any velocity and m_2 is at rest. No matter how much mass m_1 has and how fast it moves, Formulas 13 and 14 always predict the same outcome. To wit, m_1 will stop moving and m_2 will leave the collision traveling at the same velocity m_1 once had. This may sound familiar; it describes the one for one, two for two, and so on ball exchanges that occur with a Newton's Cradle.

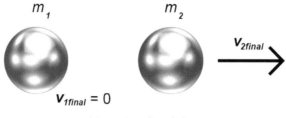

After the Collision

Using the basic physics taught in many introductory physics courses, we can state the following.

1. As m_1 strikes m_2, both objects deform slightly and for a very brief time. This causes a clicking sound as well as making each object heat up a touch; the molecules within m_1 and m_2 vibrate just a little bit faster than they did before.
2. If we add up the momentum of every vibrating molecule *within* the objects before the collision begins, the mathematical sum is zero, no overall or net momentum. This is because momentum is a vector quantity, a fact that allows for one molecule's motion in one direction to get nullified mathematically by a twin moving in the opposite one.
3. Adding more random molecular motion, as must occur when any object gains thermal energy, changes nothing. As each individual molecule vibrates faster and faster, the overall momentum *within* the objects remains unchanged at none.
4. In summary, a tiny bit of kinetic energy leaves m_1 and drops down to the molecular level *within* both objects. Moreover, a small amount of kinetic energy is responsible for the annoying clicking sound. And with a reduction in kinetic energy, there must be a corresponding reduction in macroscopic momentum. Remember that momentum relates to kinetic energy by the $^1/_2$ factor ($m\mathbf{v} \times ^v/_2 = \frac{1}{2}m\mathbf{v}^2$). That miniscule amount of momentum and kinetic energy cannot therefore be transferred as macroscopic momentum and kinetic energy to m_2. This means m_2 cannot leave the collision traveling as quickly as m_1 approached it. Therefore, momentum cannot be a conserved quantity.

Physicists may not accept some of these points, especially the last one. Some will insist that when the collision occurs, m_1 does not come to a complete stop. And on that point, they are correct. When m_1's remaining momentum is added to the momentum m_2 gets, physicists assume it means the overall macroscopic momentum remains unchanged.

As a quick and related aside, science teachers unknowingly lie to their students. And by that it is meant that the standard formulas (Formulas 13 and 14 above), or any other versions whether real or imagined, are incomplete. They do not indicate m_1 still moves after an elastic collision with another equally massive object.

Many physicists may point out that it is possible for momentum to remain unchanged despite a reduction in kinetic energy. This is a known paradox and one that gets resolved by reading this book in its entirety. For those interested in seeing the paradox in action, check out Appendix E, "Momentum and Kinetic Energy." Be forewarned, it uses mathematics to illustrate the paradox; it provides no other information.

Believe it or not, no physicist has ever precisely measured the amount of momentum before and after any elastic collision; they only think they have. In the experiments physicists have conducted over the past few centuries, a miniscule amount of momentum vanishes. It is extremely difficult to measure precisely given the way in which those experiments were conducted. Physicists simply assume momentum does not vanish because they were taught to believe this is what happens.

This scenario will not be enough to overturn more than three centuries of conventional wisdom. This makes it necessary to illustrate an even more revealing one, this time using inelastic collisions.

Scenario 2, An Inelastic Collision

Take the case of an automobile moving at 60 mph. Once the driver steps on the brake pedal, the vehicle begins slowing down. When it comes to rest, the car moves in perfect unison with the Earth. This describes the key aspect of all inelastic collisions; two objects that were once moving at different velocities combine in some fashion, and then move as if they were a single object.

Physicists assume the car's momentum gets transferred to the ground. This is then supposed to cause an extremely tiny change in the Earth's rotation. Due to the incredibly large difference in mass between the Earth and the car, this assumption has never been confirmed; no one has even tried to see if it is true. Imagine trying to measure what would be an incredibly infinitesimally small change in the Earth's rotation when a single car stops moving. At the same time, other vehicles in other locations are both slowing down and speeding up.

During the act of braking, a tiny portion of the car's momentum and kinetic energy gets transferred to the air and elsewhere. Physicists know this occurs and it is easy enough to verify. Get into any automobile, get it moving, and allow it to coast in neutral until it stops. In this simple side experiment, the vehicle begins with a quantity of momentum and kinetic energy. It ends with these completely removed due to the vehicle's interactions with the air, the road, and so on. When using the brakes, those interactions still occur but the brakes remove most of the vehicle's momentum and kinetic energy.

When the act of braking is examined with the requisite attention to detail, certain key facts stand out. Combined, they make an extremely impressive case illustrating why momentum is not a conserved quantity.

A. The typical braking system found on an automobile consist of a rotating metal surface that rubs against a stationary brake pad. This causes the molecules of one surface to rub against the molecules of a different one. This results in the rotating surface and the attached tires slowing down which, acts to stop the vehicle from moving. In short, kinetic energy transforms into thermal as one set of molecules acts on another causing all to vibrate faster and faster.
B. To help illustrate how momentum vanishes, consider the following graphic. It shows two representative molecules (m_a and m_b) within the braking components. It shows the period when they are vibrating towards one another. In terms of momentum, the equation that applies is $m_a \mathbf{v}_a + m_b \mathbf{v}_b = 0$. (Chapter 4 shows the math of how $m_a \mathbf{v}_a$ added to $m_b \mathbf{v}_b$ can equal a zero value. It does this in a way even a high school freshman can understand.)

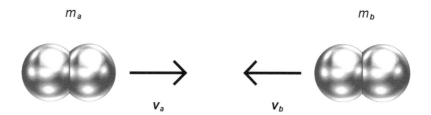

Two Molecules Vibrating Towards One Another

C. Before using the brakes, the sum of the momenta for all the individual molecules within the braking components adds up to a numerical value of zero. This is because the mass of two identical molecules is always the same and they move at the same speed but in opposite directions.

D. The next graphic shows the two molecules a short time later vibrating away from one another. Not shown are the other molecules that trap them within their assigned locations. This prevents the molecules from doing anything except vibrating in place. In terms of momentum, the equation $m_a v_a + m_b v_b = 0$ still applies but with certain modifications. To wit, the directions of v_a and v_b are the opposite of what they were previously. Moreover, the speed at which the molecules vibrate can and do change depending on various factors.

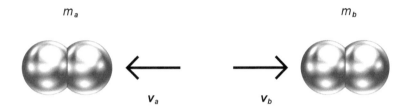

Two Molecules Vibrating Away from One Another

E. Adding additional random molecular motion, as occurs when the brakes gain thermal energy, changes nothing. The molecules vibrate faster and faster due to the braking action, but the overall sum of their molecular momentum remains mathematically the same at none. This means that the car's momentum that makes its way down to the molecular level mathematically vanishes.

F. Remember, kinetic energy and momentum use the same two variables (*m* for mass and *v* for velocity). So, if a *significant* quantity of kinetic energy transforms into thermal energy, a large measure of momentum must go along for the ride. And at the molecular level, that momentum mathematically vanishes.

Please know that points A through F represent a simplified version of reality to make a point.

As revealing as the car example is, Scenario 2 has a simpler version and one that can easily be verified experimentally. Consider an ordinary dart-board hanging on a wall. Obviously, it has no momentum since it is not moving. Now add someone throwing a dart at the target. After the dart leaves their hand, it has momentum; it has mass and moves at a measurable velocity. Once the dart hits the dart-board, it stops moving; it no longer has any momentum. We begin with two objects, one that has momentum and the other that does not. We end with a situation in which nothing moves other than those air molecules affected by the dart before it reached the target. In short, most of the dart's original momentum has vanished from the universe. Therefore, momentum cannot be a conserved quantity.

Anyone who believes the dart can affect the Earth's rotation should do the following. Find an ordinary pickup truck and attach the dart-board to its rear bumper. Attach a mirror to the front bumper and shine a laser so that the light reflects onto a surface nearby. Make sure that vehicle sits on a level surface, is not in gear, and that the brakes are not set. Now throw the same dart again. If the truck does not move as evidenced by the laser continuing to focus on the same spot, a dart thrown into a wall-mounted dart-board has absolutely no chance of affecting the Earth's rotation.

A different experiment is up next but first, a few points readers should keep in mind before reviewing Scenario 3.

Perfect Objects

When discussing inelastic collisions, science teachers use real objects. However, when elastic collisions are the topic, they use perfect or ideal objects in their lessons. This amounts to fairy tale science; perfect objects do not exist in any way, shape, form, or size. The use of perfect objects was introduced more than 300 years ago, long before anyone understood energy. It has remained in vogue ever since.

Imagine one perfect object colliding with another. Because perfect objects do not deform or are otherwise physically affected by other objects, an elastic collision would need to occur instantaneously which, is impossible. When real objects collide, they are always in physical contact with one another for a brief time. During this period, force acts through a distance due to the deformation occurring within the objects. If force cannot act through a distance because the objects do not deform, no work can be done and thus there can be no exchange of kinetic energy. This works out (pun intended) to force × zero distance equals no change in kinetic energy and so, the impossibility of perfect objects. And yes, even the steel balls in a Newton's Cradle deform when they collide. They do not deform by much but, it is enough to make them vibrate briefly. This is the reason for the annoying clicking sound and the slight thermal increase that occurs within the steel balls.

Factually, predicting the outcome of elastic collisions requires more information than just mass and velocity. The materials used also play a role as does shape and density. The solid steel balls used in a typical Newton's Cradle only approximate the predictions given by Formulas 13 and 14. This is because the properties of solid steel balls tend to *approach* those found in perfect objects if such objects could exist. Other materials have different properties and thus they react differently. For example, glass is liable to shatter, crack, or chip. Objects made from other materials can deform permanently. And when that happens, the result is a less profound bouncing effect. In other words, the predictions the usual equations produce do not happen. This means the collision problems some science teachers have their students do are a waste of time. These amount to mathematical exercises, not valid physics problems. In all fairness, the fault does not belong to today's teachers or to those that taught them. Science teachers and physics professors only pass on the same lessons they once learned.

Companion Laws

A *companion law* allows for a different law to function or exist. It is basically a situation where one physics principle cannot exist without another one playing a role. For example, bar magnets have two poles. One can be called south-seeking and the other north-seeking. When used in a compass, one end points to the north and the other to the south. These two poles are intimate companions; one cannot exist without the other. This becomes obvious when cutting a bar magnet in half; each half becomes its own bar magnet with its own north and south seeking poles. In short, if one physics principle depends on another for any reason, it is a *companion law* situation and there is nothing more to it than understanding that fact.

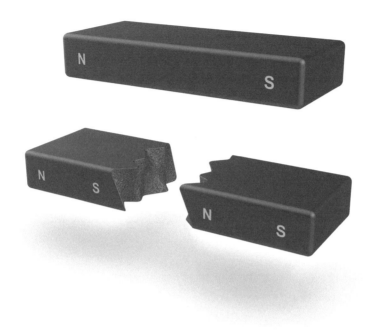

Bar Magnets

Experimental Confirmation

The ideal scientific experiment tests a prediction directly without any complications. Unfortunately, this is not always possible given that the laws of nature tend to depend upon one another. This means companion laws can sometimes complicate matters when trying to unlock the nature of nature. To get around this issue, it is sometimes necessary to do different types of experiments. Sometimes, it is only necessary to do the same experiment multiple times varying a different or the same variable each time. Unlocking the nature of nature is not easy. On the other hand, learning the known laws of physics only requires using the techniques given in Appendix A.

As for testing momentum, we live on Earth where such things as air, gravity, and the planet's magnetic field can complicate matters. Experiments other than the one described earlier might require access to highly specialized and expensive equipment. These could include such things as vacuum chambers and electronic equipment capable of measuring velocity to an extremely high degree of precision.

Surprisingly, such expensive lab equipment may not be necessary when testing momentum's conserved status; there are other options. One is to use objects made of different materials. This could be as simple as replacing the steel spheres used in a typical Newton's Cradle with spheres made from different materials and comparing their performance. A different option is an ingenious experiment any high school science class can do, and it is described in the next section.

Scenario 3, The Two-Step Experiment

Physicists have every right to demand experimental proof and so, the author offers a second measurable experiment. It is a surprisingly simple one that only requires a run-of-the-mill Newton's Cradle, a stopwatch, a sheet of graph paper, and a video camera. These items can be found in any high school. Nowadays, most students carry two of them everywhere they go; just check their cell phones.

Step 1

Place the graph paper behind the Newton's Cradle to provide reference points from which to gauge motion. Remove all but one ball from the cradle, pull this lone ball back 25 millimeters (about an inch), and release it. Time how long this single ball takes to stop moving. Record the event in such a way as to include both the timer and graph paper in the recording.

Step 2

Once again, record the event with the timer visible while the graph paper remains in place. Pull a single ball back the same distance as was done during the previous step. Release it thereby allowing it to swing back and strike a single ball; the other three balls are not used. Time how long it takes for the collisions to stop and how long it takes for all motion to cease; these times will be different.

Step 1 illustrates that a single ball takes quite some time to stop moving. This demonstrates that kinetic energy and momentum are removed by the ball's interactions with their supporting strings and the air.

In Step 2 the time recorded for the balls to stop moving is always less; the time the balls collide is far less than that. Interestingly, after the collisions cease in Step 2, we have a Step 1 situation where momentum and kinetic energy get removed primarily due to the air and the supporting strings. This tends to extend the overall time recorded in Step 2 but, it will be less than the total time recorded in Step 1.

For anyone still unconvinced about momentum's non-conserved status, do the following. Simply measure the amount of motion in Step 2 once the collisions cease. This is where the attached graph paper and video comes into play. These, in concert with a timing device, provide the means to judge the amount of motion. Next, see how much motion exists in Step 1 at the same elapsed time. In Step 1, motion diminishes due to the air and the supporting strings. In Step 2, motion diminishes for those same reasons plus the fact collisions occur. So, if there is less motion in Step 2, it means momentum vanishes during elastic collisions thereby confirming the analysis given in Scenario 1.

There is only one thing left to explore. It involves examining an extremely simple trick that has fooled everyone who has ever encountered it. Although the trick involves math, it is so simple that anyone can understand it. The only math the reader need be familiar with is elementary arithmetic, the adding, subtracting, and multiplying of numbers.

CHAPTER 4

Mathematical Trickery

Arguably, modern physics dates to the early 17th century and Italian scientist Galileo Galilee (1564 – 1642). This is because Galileo popularized the use of mathematics and experimentation that defines physics today. He stated that nature and mathematics were inexorably linked. Then there is the legend he dropped cannon balls of different sizes from the Leaning Tower of Pisa. Whether that legend is true or not, Galileo did conduct experiments to see if heavier objects fall faster than lighter ones; he found they do not.

Unfortunately, Galileo was restrained by the de facto scientific authorities of the times, aka the Pope and the Church of Rome. They decided what was and was not scientifically acceptable. No one was permitted to contradict the word of God or the physics of Aristotle (384 – 322 BC). To do so was to risk being burned alive at the stake for being a heretic. Despite such tyrannical conditions, Galileo soldiered on. Using his home-made telescope and pointing it towards the heavens, Galileo discovered that Jupiter had moons. He also saw that Venus had phases much like those we see of Earth's moon. Galileo realized these observations were at odds with the position the Church had of an Earth centric universe. In short, he came to agree with a book written by Polish astronomer Nicolaus Copernicus (1473 – 1543). That book, published after Copernicus's death, hypothesized a sun centered universe.

Galileo would go on to publish his own book in 1632 which, was nothing more or less than a thinly veiled attempt to circumvent the Church's restrictions. He used a simpleton named "Simplicio" to defend the Church's position of an Earth centric universe. For his tactless efforts, Galileo was put on trial for heresy. He escaped the usual horrific fate by recanting at his trial but, he spent the rest of his days under house arrest. Galileo's book, "Dialogue Concerning the Two Chief World Systems," would go on to remain banned by the Church for another hundred years.

When French scholar Rene Descartes (1564 – 1650) heard of Galileo's troubles with the Church, he was already living in Holland. He had relocated there five years earlier. Holland was a nation for which Descartes had once served as a mercenary. It was a mostly Catholic nation that also included a sizeable Protestant population. This made Holland a relatively safe haven for anyone whose views conflicted with Catholic Doctrine. It meant Descartes did not have to wear the same blinders Galileo had tried to circumvent.

Descartes was particularly skilled at mathematics, philosophy, and science. One of his many interests involved motion. He did not agree with the prevailing Aristotelian view of how objects moved and so, Descartes formulated his own understanding of motion. It was, in essence, an early version of Newton's Laws of Motion. His third law supposed motion was a conserved aspect of the universe. From there, he searched for a mathematical model that aligned with his supposition. Descartes basically started with the final step of today's scientific method. Specifically, instead of testing a prediction, he began by assuming a testable prediction was true. At the time, the scientific method was only in its formative stages. It had not yet been fully developed and, it was not universally accepted as the standard for scientific research.

In modern terms, Descartes's first mathematical expression for motion was the product of mass (m) and speed (s), $m \times s$, or ms. At the time, mass was not the scientific measure it is today; Descartes used weight. Weight can be crudely thought of as being the same as mass.

During Descartes's lifetime, precision was nonexistent. He could judge speed but only crudely so; there were no radar guns, lasers, or high-speed cameras. Weight could only be measured using the relatively non-standardized scales of the times. Time was measured crudely using sundials and hourglasses; pendulum clocks did not yet exist. These technological restrictions significantly limited experimentation. This forced Descartes to rely almost exclusively on logic, mathematics, and his own personal philosophic views.

Because Descartes was living in Holland, he had a crude but simple way to see if his mathematical formulation might be valid. This was due to the Dutch love of ice skating. If two skaters ran into each other, there were only two primary outcomes; each skater would continue moving or both would stop. In terms of collision theory, those collisions were either elastic or inelastic. This enabled Descartes to realize his first mathematical model for motion was both a resounding success and an absolute failure; it all depended on the type of collision.

About this time, Descartes happened to be tutoring the son of an acquaintance; the young student's name was Christian Huygens (1629 – 1695). Descartes discussed his research with his young pupil and the problem he was having with it. Huygens soon offered an interesting observation; he suggested that adding the idea of direction might resolve things. Descartes embraced his young student's idea. He changed his formulation from speed (how fast something moves) to velocity (how fast something moves *and* in what direction) to produce the expression *m**v*** (mass × velocity). And with that simple change, Descartes believed he had achieved his goal; *m**v*** appeared to work for both elastic and inelastic collisions. He went on to name his creation *momentum*, a word taken from Latin that had the meaning *moving power*.

Mathematical Blinders

Galileo could be credited with popularizing the use of mathematics in physics. Unfortunately, Descartes took this valuable tool and abused it so expertly that it set a horrible precedent. This may be difficult for some physicists to hear; they have an ongoing love affair with math that began in their first physics semester. Moreover, some university professors double dip; they not only teach physics but mathematics as well. Then there is an idea in popular culture and sometimes representative of reality. It manifests as physicists spending hours upon hours working on complex equations.

To illustrate the scientific necessity for not abusing mathematics, we return to momentum, this time as a case study. Consider the following relatively common occurrence during wartime; a cannon before and after firing a projectile.

Cannon Before Firing

Before the fuse ignites the gunpowder, we have a situation where nothing moves. The cannon is stationary as is the cannon ball within it. This means that this isolated zone containing only a cannon, gun powder and a cannon ball has no momentum whatsoever. What happens once the gun powder ignites?

Cannon After Firing

According to Newton's Third Law and in practice, the force generated by the gunpowder acts both on the cannon and the projectile. This results in the cannon ball being thrown out in one direction with enough velocity to kill and maim. On the other hand, the far more massive cannon recoils far slower in the opposite direction. Where no motion of any kind existed, we are now a witness to a tremendous quantity.

Consider the following statement taken from Chapter 1 on the now invalidated Law of Conservation of Momentum; "…, if some closed system of objects has no momentum, any action that later occurs within that area will not change the amount of momentum." No motion, then a great deal of added motion resulting in no net change in momentum. How can this be reconciled?

The answer requires taking a quick tour of an explosive situation where the misuse of math unveils its diabolical ability to deceive.

The Cannon Analogy

The Cannon Analogy mimics what occurs with a real cannon except it uses easy to follow values to keep the math simple. We start with no motion for two objects, each having a different amount of mass. For various reasons, a cannon always has far more mass than the projectiles it fires. In this analogy, the difference in mass is far less than usual; this keeps the math even simpler. The "projectile" has 1 kg of mass (m_1) and the "cannon" (m_2) has 4.

We begin with a situation where neither object moves and thus no momentum exists.

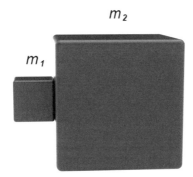

Graphic of Objects Before Force Acts

If a force (**F**) of 20 newtons acts, m_1 will accelerate to the left. That force could be due to the action of a compressed spring, gun powder, or anything else. When force acts on m_1, a force of equal magnitude acting in the opposite direction causes m_2 to accelerate to the right. Because these two objects have different amounts of mass, they accelerate at different rates. This is due entirely to the numerical relationships defined by Newton's Second Law, **F** = m × **a**. Please note that while Newton's Laws were collectively invalidated in Chapter 3, the mathematical relationship defined by Newton's Second Law remains quite valid.

In this example, one object has a mass that is four times greater than the other. This means the less massive object will accelerate four times quicker. To determine the rate at which each object accelerates, we use Formula 3 from Chapter 1. A reminder, this formula calculates the rate of acceleration when both force and mass are known.

$$a = \frac{F}{m}$$

Formula 3

$$a_1 = \frac{F}{m_1} = \frac{20}{1} = 20 \text{ m/s}^2$$

$$a_2 = \frac{F}{m_2} = \frac{20}{4} = 5 \text{ m/s}^2$$

Acceleration Calculations

Whenever force acts, it acts for the same amount of time on the two objects involved. To keep things simple, we will assume that the force acts for one full second. In many real-world situations, force acts for far less time. Witness a baseball bat striking a ball; it happens faster than anyone can blink. During this brief period, the force exerted by the bat not only changes how fast the ball moves but the direction it travels in as well.

The 1 kg object begins at rest and accelerates at 20 m/s² as determined by the acceleration calculation. This means it will be traveling at 20 m/s after 1 second (v_1). Since the 4 kg object also begins at rest but only accelerates at 5 m/s², it will be traveling at 5 m/s (v_2) in the opposite direction.

Please note that the effects of air and gravity play no role in this cannon analogy. By excluding them we keep things simple and only look at the mathematics that define momentum.

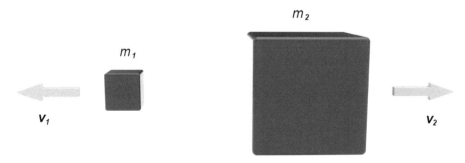

Diagram After Force Acts

From the values above, we can calculate the amount of momentum each object gets. We simply plug the values for mass and velocity into the momentum formula (p = mv) to get the following.

$$\boldsymbol{p_1} = m_1 \boldsymbol{v_1} = 1 \times (20) = 20 \text{ units of momentum}$$

$$\boldsymbol{p_2} = m_2 \boldsymbol{v_2} = 4 \times (5) = 20 \text{ units of momentum}$$

Momentum Values after Force Acts

Once we add the momentum of m_1 to m_2 we run into an issue.

$$\boldsymbol{p}_{total} = \boldsymbol{p_1} + \boldsymbol{p_2} = 20 + 20 = 40 \text{ units of momentum}$$

Total Momentum Calculation

We began with a situation in which nothing moved and thus no momentum existed, zero. After a 20 newton force acts, it appears we have 40 units of momentum. According to the Law of Conservation of Momentum, there should be none, the same as existed before firing the cannon. What is going on? The answer is that we have yet to include the idea of direction, the very thing a young Huygens noticed and Descartes embraced.

Object m_1 moves to the left while m_2 moves to the right. We can arbitrarily designate that motion to the left is negative and motion to the right is positive. Direction can also be noted in other ways but for the task at hand, the use of a positive and negative direction is the simplest method for both the layman and physicist alike. This means that m_1 moves at minus 20 m/s and m_2 moves at plus 5 m/s. Please note that the plus and the minus do not reflect how fast the objects move; they are strictly there to denote direction. So, if an object were moving at positive 2 m/s, a different object moving at negative 2 m/s would be moving just as fast; they would only be moving in opposite directions.

By adding a directional notation, we correct the previous momentum calculations thereby revealing a mathematical miracle.

$$p_1 = m_1 v_1 = 1 \times (-20) = -20 \text{ units (negative momentum)}$$

$$p_2 = m_2 v_2 = 4 \times (+5) = +20 \text{ units (positive momentum)}$$

$$p_{total} = p_1 + p_2 = (-20) + 20 = 0 \text{ units of momentum}$$

Corrected Momentum Calculations

By doing nothing more than adding a directional notation, a tremendous quantity of observable motion magically transforms into a condition of no momentum (quantity of motion). This is the exact same amount of momentum before the 20-newton force acts. This mathematical miracle explains why Descartes embraced the idea of direction. Please note that the idea that momentum means *quantity of motion* comes from Newton himself. It is the second definition he gives in his book, "The Principia".

Those who still believe momentum is a conserved quantity cannot have it both ways. If a great deal of observable motion on the large scale can mathematically vanish as just shown, it can also do this when an automobile's motion moves to the molecular level. To restate the key lesson from Scenario 2 in the previous chapter; most of the car's observable kinetic energy makes its way into the braking components. It manifests as an increase in random molecular motion, molecules that vibrate quicker than they did previously. At any given moment, there are as many molecules vibrating in one direction as there are vibrating in an opposite one. And so, the overall momentum, at the molecular level is always the same at none.

The author wins the challenge offered in the book's introduction, the bar bet if you will, without using anything other than the physicist's own understanding and some simple math.

As demonstrated, momentum can and does vanish. This fact invalidates the Law of Conservation of Momentum and by extension, Newton's Laws of Motion. For those curious about those laws and the corrections they might require, they are simple and will be given in a later chapter. Till then, the reader needs to understand a different blunder in physics; it was also made when precision was nonexistent. This requires learning a little more about kinetic energy including its unique origin story.

From Chapter 5 onwards, Newton's Laws of Motion will on occasion still be referenced and used. This is despite them being collectively invalidated by this and the previous chapter. The thing is that their use throughout the rest of this book is consistent with the proposed changes the author recommends. Not all of Newton's Laws are wrong and the proposed changes to the others are small. Sometimes a difficult problem in science, technology, or even in life only requires changing one or two small things.

CHAPTER 5

Kinetic Energy's Origin Story

Today, most everyone uses the word "energy" from time to time and properly so. Some lament about not having enough or that their kids have too much. A few older souls remember the energy crisis of the 1970s and all of us know our cell phones require periodic charging; they need electrical energy. Despite this modern prevalence and for the longest time, the concept of energy was beyond the ken of the brightest minds. Only a precious few scholars were able to approach the concept, none ever quite reaching it. Take the ancient Greeks as the prototypical example; they reasoned the universe could be broken down into four basic elements: earth, water, air, and fire. Today, we can see fire is an obvious but crude representation of energy. Centuries later, Descartes approached the concept when he formulated momentum but, he too did not quite get there. Today's modern understanding began decades later thanks to Newton's chief rival.

If the truth be known, momentum set the stage for today's kinetic energy formula. The central player in this was Huygens, Descartes's one-time student. Huygens would go on to become a celebrated scientist and mathematician in his own right. For example, he worked out how a pendulum could be used to regulate a clock. This led to the pendulum clock becoming the most accurate clock in the world until it was finally bested by 20th century atomic clocks. For this and other achievements, Huygens merited a large measure of fame. This was particularly so in Paris where he spent much of his time. There, he befriended the man who would one day become Newton's greatest rival, Gottfried Von Leibniz (1646 – 1716).

Huygens mentored Leibniz in mathematics. This included teaching Leibniz about a mathematical quantity he had been among the first to investigate. That quantity was $m\mathbf{v}^2$, and it appeared to 17th century scientists to be conserved for elastic collisions. That quantity should look familiar; it produces values that are twice as great as today's kinetic energy formula ($mv^2 = 2 \times \tfrac{1}{2}mv^2$).

As the leading scientist of the day, Newton had his share of critics. These he perceived and treated as hated enemies. Huygens was one of these albeit he was far less vocal than others and so, he was not directly on Newton's radar. Leibniz, on the other hand, was far less shy about things. He soon became Newton's archrival and the prime target for Newton's innate vindictive nature. This was Newton's defining characteristic; he did not handle criticism well and reacted poorly whenever he believed others disrespected him.

Beyond their individual accomplishments, Newton and Leibniz are known for the two boisterous disputes they had with one another. The most famous revolved around a revolutionary form of mathematics called calculus (that area of mathematics that deals with changing quantities). The other and the one of particular interest to energy began when Leibniz proposed a bold new physics concept he called *Vis Viva*. Vis Viva had the meaning of *living force* and Leibniz assigned his creation the mathematical form mv^2.

Newton took Leibniz's Vis Viva proposal as a personal attack meant to devalue him and his Three Laws of Motion. He did not believe Leibniz's bizarre idea had any scientific value. Despite his disdain for both Leibniz and Vis Viva, Newton reacted by asserting that mv would be a far better mathematical representation for Vis Viva.

Leibniz laughed at Newton's suggestion and said that if momentum were to represent Vis Viva, God would have to intervene in the universe periodically just to keep things moving. He believed the amount of living force, if defined by mv, would diminish over time. On the other hand, Newton found comfort in having the Almighty continuing to play a role in the universe. Interestingly, Newton was not shy about involving God in his research; he even praised the Almighty on multiple occasions in his book.

Even though Leibniz proposed Vis Viva, Newton's scientific stature caused many scholars to back mv over mv^2. These were predominantly English and French scientists who based their support in part on national pride; Newton was English and momentum's creator, Descartes, was French. Leibniz's biggest supporters were mainly from the Germanic speaking areas of Europe including Austria and Bavaria. As for the scholars from the other European nations, they were split between the two views.

At that time, Vis Viva did not seem to have any practical or theoretical use. Today, Vis Viva's direct descendent (energy) is the unifying factor in physics. Energy is known to change forms; for example, electrical energy can be converted into chemical energy that can later be converted into thermal and so on. Energy is the common thread that runs through all phenomena. Before any understanding of energy existed, scientists only focused on trying to understand each individual phenomenon; they did not know how to equate one with others.

As for why Leibniz thought to use mv^2 for Vis Viva, it revolved around the one problem this expression was known to have. The amounts of mv^2 before and after an inelastic collision were always wildly different. It was not a conserved quantity in such situations in the same way momentum was thought to be. Leibniz recognized this lack of traditional conservation as being the clue to something far more important. Today, we call that something the Law of Conservation of Energy.

Interestingly, the mathematics of an explosion which, is an inelastic collision unfolding in reverse, illustrates Leibniz's reasoning remarkably well. It also explains why he laughed at Newton for daring to suggest momentum could represent Vis Viva.

The Explosion Scenario

Consider a hollow metal sphere containing a measure of gunpowder. To determine the quantity of Vis Viva within only requires igniting the gunpowder and observing the results. In the simplest scenario, half the metal sphere flies off to the left and the other half goes to the right. In this situation, Leibniz would state that the quantity of Vis Viva would be equal to the Vis Viva of the half sphere moving to the left added to the amount the other half receives.

Exploding metal sphere

If each half sphere has a mass of 1 kg and they move at 10 m/s after the explosion, we can calculate how much Vis Viva exists.

$$\text{Total Vis Viva} = m_1 v_1^2 + m_2 v_2^2$$

$$= (1 \times 10 \times 10) + (1 \times 10 \times 10)$$

$$= 100 + 100$$

$$= 200 \text{ units}$$

Vis Viva Calculation

If there is less gunpowder, each half sphere would leave at a slower velocity. This would result in a calculation that would be less than 200 units. With more gunpowder, the opposite would be true; the half spheres would be traveling faster than 10 m/s. This would result in a calculation indicating a greater quantity of Vis Viva. This is as should be expected; the more gunpowder in play the greater the quantity of Vis Viva.

Please note that the 200 units of Vis Viva is technically inaccurate. Beyond the mathematical difference between today's kinetic energy and yesteryear's Vis Viva formulas, some of the Vis Viva transforms into other forms of energy (heat for example). At the time, Leibniz did not understand this; he was only looking at motion and its immediate causes.

Applying momentum to the same situation results in a numerical value of zero units of momentum. This makes absolutely no sense as a description for Vis Viva or energy.

$$P_{total} = m_1 v_1 + m_2 v_2$$

$$P_{total} = 1 \times (-10) + 1 \times (+10)$$

$$P_{total} = 0$$

Momentum Calculation

If more explosive is used, the velocities of the individual half spheres increases but, momentum's mathematical sum is still zero. Please note that the same principles that apply when the mass of the two objects are different (as previously shown in the Cannon Analogy) applies when the masses are the same. In fact, a zero value always occurs because Descartes stopped developing momentum when he saw his mathematical creation produce this result.

The Explosion Scenario amounts to an experimental test that excludes. If a proposed expression passes this test, it does not necessarily mean that expression should represent mechanical energy. This fact becomes blatantly obvious simply by looking at mv^2 and $½mv^2$. Both expressions pass this test but only one of these could represent the energy of motion. In short, the Explosion Scenario only rules out momentum.

The First Experiment

For years after its introduction, the question of how to describe Vis Viva mathematically remained unanswered; should it be mv^2 or mv? This was because there was no known experimental evidence supporting either formulation. There was also no universal understanding of the Explosion Scenario. Moreover, no one considered that Vis Viva had any scientific or practical value. Then out of the blue, Emilie Du Châtelet (1706 – 1749) took it upon herself to resolve the controversy. She was a Frenchwoman and Aristocrat who was also a bit of an enigma in that she had interests well beyond her station and time. To begin, she had a good grasp of mathematics and was up on all the latest advances in science. This was unusual for an 18th century woman, unheard of really. Du Châtelet also understood Newton's book, "The Principia", well enough to be the one who translated it from Latin into French. Moreover, she had a firm grasp on Leibniz and his philosophic views. In short, Du Châtelet seemed to have all the right qualifications to decide the issue. The only problem was her gender; it was a liability given the era in which she lived. This forced Du Châtelet to hide her skills and ambitions from all but a precious few confidants and allies.

For a time, Du Châtelet contemplated how to go about deciding the issue without success. Then, as she was looking at what others were doing or had done, a solution presented itself. It took the form of an experiment Dutch scientist and former lawyer Willem s'Gravesande (1688 – 1742) had devised. In 1715, he travelled to England on behalf of his country. While there, s'Gravesande had occasion to meet Newton and became enamored by the world's greatest scientist. His fascination for Newton was so great that he decided to abandon his legal career for a scientific one.

By modern standards, the s'Gravesande Experiment is extremely simple. His experiment consists of nothing more than dropping small metal spheres from various heights into a soft bed of clay and then measuring how deeply they sank.

Du Châtelet figured momentum would be the answer to the Vis Viva question if the relationship between the depths of penetration was linear with respect to velocity. In other words, if a doubling of velocity resulted in a doubling of the depth to which a sphere penetrated the clay, that would be a direct (linear or one for one) result favoring momentum. This did not occur; instead, she found that the depths to which the spheres sank into the clay always followed an entirely different pattern. In short, if a sphere travelled twice as fast as another identical one, the faster sphere would penetrate the clay four times as far. If the faster sphere travelled three times as fast, it would sink nine times deeper. It was a dramatic result that Du Châtelet took as proof that Leibniz's hypothesis was indeed valid. It also meant that Vis Viva must have the mathematical form he proposed ($m\mathbf{v}^2$).

Du Châtelet

Du Châtelet exploited one of the two key differences between mv and mv^2. The first is that momentum is a directional expression; this was the key to momentum's perceived success. Leibniz's mv^2, due to the rules of mathematics, produces a result that is scalar.

The second difference and the one Du Châtelet exploited was that an object's momentum relates directly to how fast an object moves. Double an objects velocity and momentum also doubles. With mv^2, it is an entirely different situation; the values this expression produces are always related to the square of the object's velocity. When velocity doubles, the numerical value of mv^2 increases by a factor of four (2 × 2). Should velocity triple, the numerical value of mv^2 increases by a factor of nine (3 × 3), and so on.

Please note that Du Châtelet was able to calculate velocity by simply measuring height and using a known formula. This told her that a quadrupling of height meant an object hit the clay twice as fast. This proved to be quite convenient given how she intended to test mv and mv^2.

Once Du Châtelet finished her experiments, she had the evidence Leibniz had failed to provide. In 1747 Du Châtelet published her findings to a scientific community that ignored both her and her conclusion.

Years later, the scientific community finally realized Leibniz's vision might have some value. This new generation of scientists began using mv^2 but not because of Du Châtelet; her experiments were never scrutinized, repeated, or considered valuable. She was a woman working in an overwhelmingly male vocation. This meant she could be ignored by men with egos and wives that knew nothing of such elite matters. This new generation of scientists accepted mv^2 because they were finally able to grasp why momentum could not represent Vis Viva.

It is worth noting that the scientific method does not contain a provision allowing a hypothesis to rise to the status of an accepted theory by default. Sherlock Holmes on the other hand might say that when you eliminate the likely, whatever is left, no matter how improbable, must be the answer. Of course, Holmes is a fictional character and so, his views on the matter should not hold much weight. Sadly, today's physicists unknowingly use his philosophic technique. They state the kinetic energy formula must be valid because momentum cannot represent energy. Evidently, it is standard practice for physics students to sleep through the lectures on the scientific method and embrace the fictional notions of men who never lived. Ironically, few physicists know of s'Gravesande or Du Châtelet; their contributions to physics remain largely untaught and unacknowledged. On the other hand, Sherlock

Holmes is known by most everyone; he has been the central fictional character in numerous movies and television programs.

Birth of Today's Understanding

As the 19th century began, English polymath Dr Thomas Young (1773 – 1829) renamed Vis Viva with his introduction of the word "energy". In his famous series of public lectures circa 1805, Young explained energy's relationship to height. He pointed out that an object that falls from a height twice as great as another identical object has a value of energy twice as great. This reference to height had the effect of aligning mv^2 with the modern understanding of force acting through a displacement. In those lectures, he did not modify Vis Viva's original mathematical form but, the stage was set. In 1829, French scientist and mechanical engineer Gaspard Gustave De Coriolis (1792 – 1843) published a Paper entitled, "On the Calculation of Mechanical Action." In it, he upgraded mv^2 to $\frac{1}{2}mv^2$; he also introduced the word "work" as it exists in the modern physics lexicon. Incidentally, De Coriolis is known today more for his ideas on rotating systems. This includes the difference in the direction of winds between the northern and southern hemispheres, aka the Coriolis Effect.

After De Coriolis, others added to the understanding of energy. One of the most notable was English physicist James Prescott Joule (1818 – 1889). He wrote a Paper in 1844 entitled, "On the Mechanical Equivalent of Heat". In it, Joule equated thermal energy with mechanical. His efforts, along with those of others dispelled the prevailing Caloric Theory for heat. In short, the work of 19th century scientists replaced the idea of a self-repellent fluid called caloric with the kinetic theory of heat. For Joules' efforts, scientists named the basic unit of energy in his honor, the "joule". Another notable individual was German physicist Albert Einstein. In late 1905, he published a Paper entitled, "Does the Inertia of a Body Depend Upon Its Energy Content". In it, Einstein equates mass with energy which gave rise to the one physics equation most people know exists, $E = MC^2$. Readers might be surprised to know that Einstein did not write this equation in this manner. For example, he used the letter "L" to represent energy instead of the letter "E".

After reading this book, readers might want to check out contemporary author David Bodanis' book entitled, "$E=MC^2$: A Biography of the World's Most Famous Equation". It has far less math than this one. It does not expose the blunders within physics. It is more about detailing the contributions of such historical figures as s'Gravesande, Du Châtelet, and numerous others that led to both the kinetic energy formula and Einstein's famous equation.

More on Work and Kinetic Energy

In Chapter 2, Formula 7A equated work with kinetic energy. That equation only deals with objects that begin at rest. If the truth be known, work can be done to objects that are already moving. This means the reader needs to understand just a tiny bit more about work and kinetic energy. This additional information will be used to expose certain other missteps of the past that remain a part of physics as she is taught today.

When an object is already in motion, the work done to it is equal to the *change in* the kinetic energy that object experiences. This is given in Formula 15 below or often stated as "work = Δ kinetic energy". The mathematical symbol "Δ" comes from the Greek alphabet. It is the letter *delta* and it can be used to mean change in.

Mathematically, Formula 15 also works for objects that begin at rest. In that case, the initial amount of kinetic energy, which is none, gets subtracted from the final amount.

$$\boldsymbol{Fd} = \frac{m\boldsymbol{v}_{final}^2}{2} - \frac{m\boldsymbol{v}_{initial}^2}{2}$$

Formula 15

where,

$\boldsymbol{v}_{final}$ refers to the velocity of the object after work is done.

$\boldsymbol{v}_{initial}$ refers to the velocity of the object before work is done.

F refers to the applied force.

d refers to the displacement force acts through.

m is the amount of mass the object has.

This formula is only ever used as written here. This means that any time an object slows down, the work done to it will be negative. For example, this occurs when a spacecraft orbiting the planet wishes to return; it must first slowdown and this requires negative work.

Irony

Irony cannot be ignored when talking about momentum and kinetic energy. Leibniz used mv^2 to describe Vis Viva based at least partly on the advice of his mentor. Huygens was also the one who suggested the addition of direction to Descartes. This resulted in momentum, an expression that cannot represent Vis Viva, or energy. The irony goes even deeper in that Descartes ruled out his first formulation (mass × speed also written as $m \times s$ or ms) for the very reason Leibniz embraced mv^2. Both ms and mv^2 are not conserved when it comes to inelastic collisions. Stated in another way, three expressions are known to pass the test called the Explosion Scenario; the first is mv^2 as well as its more modern form $½mv^2$. The third one is *mass × speed* since it has no direction associated with it; the Explosion Scenario only rules out momentum.

At this point some readers might ask why any attention is being paid to mass × speed since Du Châtelet proved Leibniz was correct. If the truth be known, she proved nothing of the kind; she merely demonstrated that she was not the ideal scientist to decide the matter. Du Châtelet made the greatest blunder any physicist could make. And when that blunder gets exposed in the next chapter, the s'Gravesande Experiment becomes the most ironic experiment in history, bar none.

CHAPTER 6

The s'Gravesande Experiment

From a modern perspective, there are many reasons to believe in the validity of the work and kinetic energy formulas. These begin with their longevity; together, they have been around in their current form for almost two hundred years. Moreover, physics students learn momentum cannot represent kinetic energy. And then there is an experiment many students do whose obvious conclusion is rather compelling. The Academic Experiment, as it is called in this book, is so impressive that no one questions it in any way. However, when the author examined it, he uncovered an issue hidden within it so insidious it must be seen to be fully appreciated. If the reader did any experiments relating gravitational potential energy with kinetic, he or she must read and understand Appendix D before continuing. To do otherwise is to do a disservice to the reader and the author alike. For those who have no knowledge about that experiment, the only thing anyone truly needs to know is that it causes students to make the same blunder Du Châtelet did. In her case, the blunder is relatively obvious; it is far less so in the Academic Experiment, especially when one is a student tasked with doing it.

A New Hypothesis

This chapter also offers a new hypothesis which, has the effect of providing a far more viable option to describe energy than momentum ever was. In formal terms, the author proposes that the product of mass (m) and speed (s) might be a better description of the energy a moving object has. With no references to direction, mass × speed easily passes the test that is the Explosion Scenario.

Some physicists may object to mass × speed because it contains no vectors. The thing is that Descartes only added a directional component to his original formula because it appeared to validate his supposition. Unfortunately, that supposition was rendered impotent by Chapters 3 and 4. Then there is the fact physicists already exclude direction when talking about light; they talk about the *speed* of light, not its velocity. To this we can add the hypocrisy of the kinetic energy formula itself. It may use vectors in its formula but, its answers are scalars. In short, no one should assume mass × speed is an invalid option simply because it contains no vectors.

Equation 16 represents the New Hypothesis's basic form. This equation describes the situation where an object begins at rest; Equation 17 applies to objects that are already in motion.

$$Ft = ms$$

Equation 16

$$Ft = \Delta ms$$

Equation 17

where,

> *F* is the amount of force that acts in the same direction that an object changes its motion. Direction is not otherwise noted; hence the variable is no longer written in bold type. The variable "*F*" represents the magnitude of force and is given in newtons. The definition for the newton does not change; one newton is still the amount of force that accelerates a one-kilogram object at a rate of 1 m/s^2.
>
> *t* is the time force acts and its basic unit is the second.
>
> *m* is the mass of the object given in kilograms.
>
> *s* is the speed of the object, and its basic unit is meters per second.
>
> *Δ* is the Greek letter delta having the meaning change in.

If we added directional components to *Ft* and *ms*, they would be called impulse (***F**t*) and momentum (*m**v***). The name for the nondirectional version of impulse is *effort* (force × time). The name for the non-directional version of momentum is *dynamic energy* (mass × speed). From the Explosion Scenario, we learned that momentum cannot represent mechanical energy. Its non-directional version might; it all depends on whether experimental evidence confirms the New Hypothesis.

The New Hypothesis carries with it a unique prediction that describes energy's relationship with speed. This means this hypothesis passes the first test required by the scientific method; it has a testable prediction. Just as importantly, its prediction differs from the one given by the kinetic energy formula. With dynamic energy, a doubling of speed translates to a doubling of *ms*; a trebling of speed produces a value of *ms* that is three times as great. This is a one for one, a direct relationship between the speed of an object and its energy content. On the other hand, the kinetic energy formula produces values that grow at an ever-changing rate with respect to an object's velocity. This difference in how energy increases with how fast an object moves should make it relatively easy to determine which formula is in accord with reality.

The s'Gravesande Experiment

The only experiment done before scientists elevated Leibniz's hypothesis to an accepted theory was the s'Gravesande Experiment. Du Châtelet used it to test both *mv* and *mv²*; she concluded Leibniz's choice was the correct one. Unfortunately, no one repeated her experiment and so, no one checked her work for any errors, obvious or otherwise.

The s'Gravesande Experiment is one in which several identical metal spheres fall from various heights into a soft bed of clay. The idea is to see how far the spheres penetrate. In modern terms, the premise is that as an object slows down, it loses the energy associated with motion. And if energy gets removed at a steady rate, that action reveals the relationship energy has with respect to velocity, or speed. This experiment does not require knowing how much energy gets removed, only that it occurs at a uniform rate. In practical terms, this is simply a situation where the rate of deceleration is a constant. Stated differently, the force acting on the object is a constant throughout the experiment.

It does not require much imagination or even ingenuity to devise similar experiments to the one Du Châtelet conducted. For example, drive a car to various speeds and slam on the brakes. See how far the vehicle travels before it stops. Find a nice flat putting green, hit golf balls to various speeds, and see how far they travel before stopping. The common factor in these versions and the original s'Gravesande Experiment is that something slows down at an assumed uniform rate. And when objects slow down, two things are always in flux (changing); the object travels a certain *distance,* and this occurs over *time*.

The problem with the s'Gravesande Experiment Du Châtelet conducted goes to when the spheres reach the clay; the spheres quickly stop moving. In fact, they stop moving so fast that it was impossible for her to notice time varied. To Du Châtelet, the spheres stopped instantaneously or nearly so. To make matters worse, even if she had thought about time, Du Châtelet did not have the means to measure time to any degree of precision. This is especially so since the times involved were mere fractions of a second.

In the final analysis, Du Châtelet only measured the distance the spheres penetrated the clay. She completely missed time as a variable. This means Du Châtelet's conclusion is utterly worthless; it is based on half the data a properly conducted experiment produces. And with this revelation, the kinetic energy formula no longer has any science behind it. None of the reasons given at the beginning of this chapter align with the scientific method. This even includes the Academic Experiment some science teachers and physics professors use. It is also improper to cite experiments or conclusions that depend on the validity of the kinetic energy formula itself; these amount to self-fulfilling arguments, not science. The kinetic energy's only claim rests on its longevity which, is no claim at all. Its only hope is that the inclusion of the time variable does not alter Du Châtelet's conclusion.

Due to the nature of a s'Gravesande type experiment, it is not difficult to calculate a set of representative values. These are the distance an object travels during the act of slowing down and how much time that takes. This only requires knowing the rate at which an object slows down and using two simple formulas. The calculated values will be the same as those that occur experimentally. If there are any numerical discrepancies, they will be small and in keeping with the expected margin of error that occurs whenever measurements are taken.

If the reader is uncomfortable with the basic formulas involving distance, time, speed, and acceleration, they may skip down to the "Experimental Data" section below if they wish. However, the following is not difficult and should already be known by anyone who went to high school. Despite this, the mathematics is still explained in simple terms with examples. If anyone has trouble with any of this, blame your high school math teachers.

The following variables are the only ones that have any use or value when looking at an s'Gravesande type experiment. None are vectors because direction plays absolutely no role in the outcome; thus, direction can be safely ignored.

a is the rate at which deceleration takes place in meters per second per second, an unchanging value. In this exercise, the rate of deceleration is specified. In any real experiment, this value could be found using the recorded data of time and distance.

s is the initial speed of the object in meters per second just before it reaches something that will remove energy at a uniform rate. This value is at the discretion of the person doing the experiment and should encompass numerous different values for reasons that will soon become obvious.

t is the time in seconds the object takes to stop moving once deceleration begins. When the experiment is done, this value would be measured instead of being calculated.

d is the distance in meters the object travels while decelerating. When the experiment is done, this value would be measured instead of being calculated.

Time

We know that the time an object accelerates equates to its final speed. For example, if an object accelerates at a rate of 5 m/s² and it begins at rest, it will be moving at 5 m/s after one second (1 second × 5 m/s² = 5 meters per second). If it accelerates for two seconds, it will be moving at 10 m/s at the end of that time (2 seconds × 5 m/s² = 10 meters per second). This translates to a useful formula, *s = t × a* (speed is equal to the product of time and acceleration). When we rearrange this formula to find *time* instead using the simplest ideas in algebra, we get Formula 18. This formula works quite well when the goal is to calculate the time an object takes to stop moving. It only requires knowing the object's initial speed (*s*) and the rate at which the object decelerates (*a*).

$$t = \frac{s}{a}$$

Formula 18, the time an object takes to stop moving

Distance

If you are driving at 30 miles per hour, the distance you travel is the product of that speed and time (distance = speed × time). So, if you travel at 30 mph for two hours, you will have driven 60 miles (30 mph × 2 hours = 60 miles). However, in the s'Gravesande Experiment, speed varies at a known and uniform rate. An object starts off at some initial speed and its journey is over when the object no longer moves. Since the experiment relies on an unchanging rate of deceleration, we can easily calculate the object's average speed. This only requires dividing the initial speed by 2. The distance (*d*) an object travels while it decelerates will then be equal to the time (*t*) it moves at its average speed.

$$d = t \times \frac{s}{2}$$

Formula 19, the distance an object travels

The variable "*s*" is the object's initial speed, and the variable "*t*" is the time it takes to stop moving. This means it is necessary to calculate time first.

Experimental Data

Table 1 shows the values for time and distance when an object slows down at the rate of 2 m/s^2. The table uses a set of initial speeds ranging from 2 to 20 m/s. To keep the math even simpler, the increase in initial speeds always jumps by 2 m/s. If we were to jump by 1 m/s, some of the calculated values would be fractions. Remember, the goal is not to complicate things; it is to keep things as simple and informative as possible.

Initial Speed	Time	Distance
in meters per second	in seconds	in meters
2	1	1
4	2	4
6	3	9
8	4	16
10	5	25
12	6	36
14	7	49
16	8	64
18	9	81
20	10	100

Table 1

From Table 1, it should be quite apparent that both the initial speed and the time in seconds for an object to stop moving change at the same rate. It is in other words, a one for one relationship. Double the initial speed from 2 to 4 m/s and the time to stop also doubles. Treble the initial speed from 2 to 6 m/s and the time to stop follows suit; it trebles. And this pattern holds good throughout. Double the initial speed from 10 to 20 m/s and time still doubles; it goes from 5 to 10 seconds.

As for distance, it increases at an entirely different rate. When the initial speed increases from 2 to 4 m/s (a doubling of speed), distance increases from 1 to 4 meters, a four-fold increase. When initial speed trebles, distance increases by 9 times. And this pattern also holds good throughout. Double the initial speed from 10 to 20 m/s and distance once again increases by a factor of four; it goes from 25 to 100 meters. This is an ever-increasing rate of change in distance that relates to the *square* of the changes in initial speed. So, if speed doubles (× 2), the distance increases four-fold (2 × 2). If speed trebles (× 3), distance increases by a factor of nine (3 × 3) and so on.

When plotted on a graph, we get a visual representation of how the time to stop moving and the distance traveled relate to initial speed. In Graph A, time is a nice straight line and that indicates a one for one relationship. Distance is a smoothly curving line whose numerical values increase at a faster rate than initial speed.

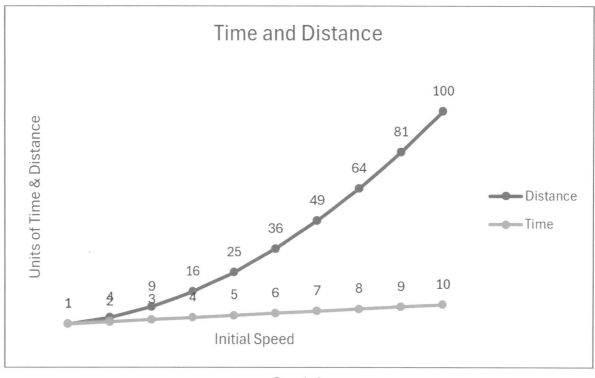

Graph A

What happens if we only measure or look at one of the variables from the s'Gravesande Experiment? The answer is that it becomes possible to make the case for the kinetic energy formula *or* the product of mass and speed; it all depends on which variable we focus on. If we only look at the time an object takes to stop moving, a one for one relationship is the result and that aligns with the New Hypothesis (the straight line). If we do as Du Châtelet did and only consider the distance objects travel while decelerating, we wind up with the same conclusion she made; it becomes "proof" for the kinetic energy formula (the curving line). Whichever one (time or distance) we *arbitrarily* select, that choice will naturally favor a particular outcome. This, in essence, amounts to a self-fulfilling prediction that is argumentative, not science.

The proper thing to do is to include both time and distance when analyzing an s'Gravesande type experiment. To exclude either variable requires a compelling reason and there are none for either time or distance. Both variables change during the experiment and so we are left with combining the *two* sets of data. The only logical way to do this is to divide the distance traveled while slowing down by the time it takes ($d \div t$).

Table 2 shows that when the initial speed doubles, so too does the distance ÷ time. If the initial speed trebles, so too does the distance ÷ time. When these experimental results get mapped on Graph B, the result is a nice straight line. This aligns with the New Hypothesis and indicates the product of mass and speed should replace the kinetic energy formula.

Initial Speed	Distance ÷ Time
In meters per second	In meters per second
2	1
4	2
6	3
8	4
10	5
12	6
14	7
16	8
18	9
20	10

Table 2

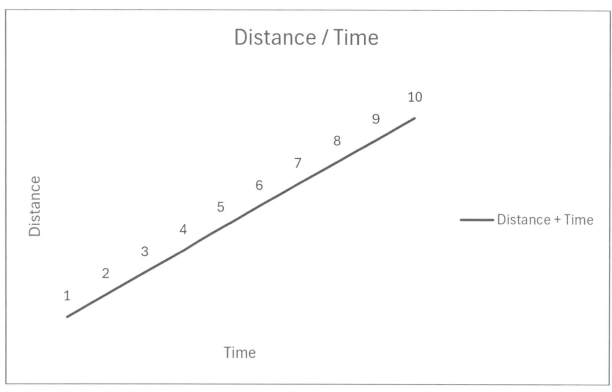

Graph B

As mentioned in the previous chapter, the s'Gravesande Experiment is the most ironic experiment in history. It was first used to show Leibniz's choice (mv^2) for Vis Viva was correct. However, when the experiment is performed properly and the analysis includes the changes that occur to both time and distance, it produces an entirely different conclusion.

Many physicists are liable to balk when seeing Graph B if for no other reason than it contradicts their education. If this experiment supported the kinetic energy formula, they would instead be praising it. Since this is not the case, many physicists will seek to locate any excuse they can to invalidate the experiment. Some physicists will be quick to point out that the values on Graph B represent average speed as if that were a monumental revelation, or that it somehow invalidates the analysis; it does not. The results align perfectly with the prediction given by the New Hypothesis, a one for one relationship.

Regardless of the significance of average speed, the best way to analyze experiments of this character is by thinking of the experiment as a "black box"; data enters and data exits. In this case, the input is a linear change in the initial speed; the output is just as linear. Neither Leibniz's initial formulation (mv^2) nor its successor ($\frac{1}{2}mv^2$) produce values that are linear with respect to changes in initial speed, or velocity.

Actual Experiments

Please note that in producing a data set using mathematics, nice round numbers resulted. When conducting actual experiments, the data is unlikely to turn out as nice even if the rate of deceleration is the same; the values are apt to look like 1.98 seconds and 4.05 meters instead of 2 seconds and 4 meters. There is always some margin of error when measuring experimental data. However, the data will graph very much like those shown here with one slight caveat.

Take the case of a truck traveling at 100 mph and another identical one traveling at 10. If we apply the brakes exactly the same for both, the faster vehicle will take less time to reduce its speed by 10 mph. This is because the faster truck encounters far more air. The more air a vehicle must push out of the way, the greater the counterforce air provides which adds to the counterforce the brakes themselves deliver.

This means s'Gravesande type experiments can have complicating issues. In those cases where the additional unwanted counterforce is significant, it will produce values that land on a graph in a nonlinear way but, only at high initial speeds. Despite such hiccups, s'Gravesande type experiments easily support the New Hypothesis. Moreover, those issues completely vanish when the additional counterforces are taken into consideration.

Once again and for the record, please note that the author is not asking anyone to take his word on anything; he wants others to verify everything. This also includes examining the experiments mentioned in the two upcoming chapters. One of those is so incredibly simple and important that the author must repeat a statement made earlier. "The profound is not complicated and hard to understand; it is always simple and easy to grasp."

CHAPTER 7

The Cart Experiment

Instead of analyzing an object as it slows down, the Cart Experiment examines the situation in reverse. It does this by accelerating an object on two occasions using the same amount of energy each time.

This chapter also includes the details of an entirely different experiment that uses readily available equipment. Its simplicity is only exceeded by its effectiveness. It clearly illustrates why the kinetic energy formula should be abandoned. And if it were not for the fact blinders do in fact exist, it would be the only experiment required to invalidate the work and kinetic energy formulas.

Phase 1

Phase 1 exists primarily to establish a baseline velocity as well as ensuring no issues with the equipment. If Phase 1 is repeated several times and the projectile's velocity does not vary, one can confidently state that the equipment always applies the same amount of energy.

In the graphic shown below, the cart's initial velocity is 0 m/s. According to Newton's Third Law and in practice, as the cart fires the projectile, the cart reacts; it will move in a direction opposite to that of the projectile. The projectile's final velocity or speed relative to the radar gun is 5 m/s. That radar gun is at rest with respect to the ground and serves as the required reference point for all velocity measurements.

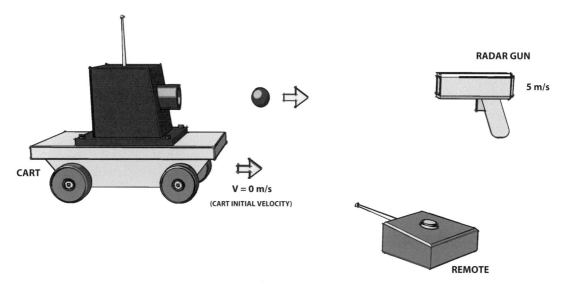

Phase 1

The value of a reference point for velocity or speed is shockingly easy to understand if you have ever been in an automobile. The speedometer tells the driver how fast he moves with respect to the road. If the posted speed limit is 65 mph and the speedometer indicates you are moving at 64, you know there is no danger of getting a speeding ticket. If another vehicle's speedometer reads 84 mph and it is moving in the same direction, it can be said to be moving at 20 mph with respect to your car (84 – 64 = 20 mph). In that situation, the reference point for his 20-mph speed is your vehicle, not the road. This means that if nothing changes, the faster vehicle will be 20 miles further down the road after one hour's time. If he is caught by the police, he cannot claim innocence by citing your vehicle as the reference point; it is the wrong one for the situation.

If the truth be known, any mention of velocity or speed to be valid must include a proper reference point or, it should be obvious from context. News outlets tend to violate this when they report a spacecraft docking with the International Space Station (ISS). They point out that the ISS moves at about 17,000 mph. And while that is true with respect to the ground, the fact is that when a spacecraft is about to join up with the space station, the speeds involved during the docking procedure mimic those of a person walking up to and getting into a parked car. So, by referencing the fantastic speed of the ISS, news outlets mislead. The docking situation may be dangerous but by using the 17,000-mph value, they seek to make it even more so in the eyes of their audience.

The cart is any four wheeled vehicle capable of supporting a mechanism that fires a reusable projectile. The mechanism could be practically anything from a rubber band to an electrically operated arm. The only thing the mechanism need be is consistent; it must produce the same result every time. In this way, the energy applied to the projectile never changes; it is always the same amount.

When the mechanism fires a projectile, the cart reacts; it begins at rest and will then move to the left. Its final velocity is wholly dependent on its mass (mechanism plus cart) and the force that acts.

Kinetic Energy Prediction

In Phase 1, the projectile achieves a velocity of 5 m/s as calculated by the formula below.

Please note that as always, the chosen values keep the math simple; they were not selected because they support the author's hypothesis. The thing is that real world values make things unnecessarily difficult; it is far easier to multiply 2 by 4 than 1.98 by 4.05.

$$\boldsymbol{Fd} = \frac{mv^2}{2}$$

Formula 7A from Chapter 2

$$\boldsymbol{Fd} = \frac{2 \times 5^2}{2} = 25 \text{ joules}$$

Calculating work (force × distance) done to get a 2 kg object moving at 5 m/s

In Phase 2, the mechanism is reset and made to move at 5 m/s. Thereafter, the mechanism acts on the same projectile and, as before, it does 25 joules of work to the projectile. After a velocity measurement is recorded, we compare it to the predicted final velocity (v_{final}) using Formula 15.

Physicists take note; the author fully understands any objections they may wish to voice at this time. Those concerns will be addressed in two different ways, first below and later in another chapter. Until then, please hold off on any judgements; remember, the kinetic energy formula failed in the s'Gravesande Experiment. Moreover, if anyone wishes to cite the relationship of height with the kinetic energy formula, the author encourages them to read Appendix D first.

$$Fd = \frac{mv_{final}^2}{2} - \frac{mv_{initial}^2}{2}$$

Formula 15 from Chapter 5

$$v_{final} = \sqrt{\frac{2Fd}{m} + v_{initial}^2}$$

Formula 15 Rearranged to solve for final velocity

$$v_{final} = \sqrt{\frac{2 \times 25}{2} + 5^2} = \sqrt{50} = 7.071 \frac{m}{s} = 7.071 \text{ m/s}$$

Kinetic Energy Prediction for Phase 2

If the projectile's velocity after Phase 2 is about 7 m/s, that result would be enough to confirm the kinetic energy formula should represent mechanical energy.

Dynamic Energy Prediction

To determine how much effort (*F* × *t*) the cart's mechanism applies requires using Equation 16 from Chapter 6. One need only plug in the projectile's mass (2 kg) and its speed (5 m/s) at the end of Phase 1.

$$Ft = ms$$

Equation 16

$$Ft = 2 \times 5 = 10 \text{ units}$$

Calculating how much *effort* the cart's mechanism produces.

The New Hypothesis predicts an entirely different final speed (s_{final}) for the projectile at the conclusion of Phase 2. This begins with an expanded version of Equation 17 ($Ft = \Delta ms$) rearranged to solve for the projectile's final speed. This is very much like rearranging Formula 15 except the New Hypothesis's formula is a bit simpler. Thereafter, we plug in the 10 units for effort just calculated, the object's mass of 2 kg, and the initial speed ($s_{initial}$) of 5 m/s for the projectile as Phase 2 begins.

$$Ft = ms_{final} - ms_{initial}$$

Equation 17 Expanded

$$s_{final} = \frac{Ft + ms_{initial}}{m}$$

Equation 17 Expanded rearranged to solve for final speed

$$s_{final} = \frac{10 + (2 \times 5)}{2} = 10 \text{ m/s}$$

Dynamic Energy Prediction for Phase 2

If the projectile's velocity after Phase 2 is 10 m/s, this would be enough to confirm the corrected conclusion of the s'Gravesande Experiment.

Phase 2

Phase 2 begins by returning the projectile to the cart and resetting the mechanism. Thereafter we get the cart moving at the 5 m/s velocity recorded in the previous phase. This means that the sphere within the cart's mechanism is also moving at 5 m/s.

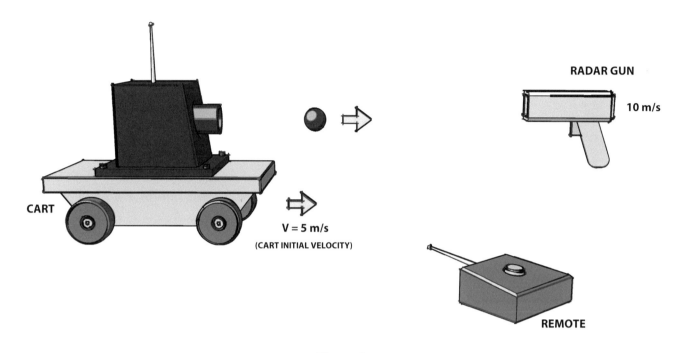

Phase 2

When the cart fires the sphere in Phase 2, the radar gun records a velocity of 10 m/s. As before, the cart reacts but, in Phase 2, it slows down.

The measured 10 m/s velocity matches the prediction given by the New Hypothesis. This confirms the s'Gravesande Experiment's corrected conclusion.

Arguments Against the Cart Experiment

Any competent physicist will admit the projectile's velocity after Phase 2 is 10 m/s. This will not deter some from disagreeing with the experiment's obvious conclusion. Some may focus on the fact that it took energy to get the cart and the projectile up to the 5 m/s initial velocity for the start of Phase 2. When they do the math that includes the cart, their numbers work out perfectly. Those calculations are not discussed in this chapter; they will be examined in a later one. Until then, the following should be sufficient to handle any objections to the Cart Experiment. It exposes the undefendable double standard physicists were taught and are now using.

Consider someone lifting a box and placing it on the first step of a flight of stairs. The person then goes up one step himself and then lifts the same box up to the second step using the exact same motions as before. A science teacher or physics professor, if asked about this, would state that a doubling of work has occurred *to* the box. It took a specific quantity of work to raise the box the first 20 centimeters for a single step and an identical amount of work to raise the box another 20. These educators would go on to state that the work *done to the box* if raised 40 centimeters in one motion is no different than the work done to it raising it half that distance on two separate occasions. They would then say that the energy used by the person to climb the first step is irrelevant; it was merely the price paid to elevate the box in two stages for the express purpose of using the exact same motions in each. Most importantly, they will only focus on the work *done to* the box.

In Phase 1 of the Cart Experiment, the cart's mechanism accelerates the projectile from rest to 5 m/s. After returning the projectile to the cart and resetting the mechanism, they are made to travel at that same speed. This is like climbing a step for the express purpose of using the same motions in the exact same manner as was done before. When the mechanism fires the projectile during Phase 2, it uses the exact same amount of energy again. This results in the projectile's speed doubling from 5 to 10 m/s. It took 10 units of effort to get the projectile from rest to 5 m/s and another 10 units applied to the same projectile to double that speed.

Quite notably, if 20 units of effort were to be applied when the object was at rest, the projectile would wind up traveling at 10 m/s, a speed twice as fast as occurs during Phase 1. Every physicist knows this to be true because they have done problems involving impulse and momentum. Remove direction from impulse and we have effort (force × time); do the same for momentum and the result is dynamic energy (mass × speed).

A few physicists may rail against the Cart Experiment by blindly parroting the lessons they learned in high school. They will state that to double the projectile's velocity, it is necessary to quadruple the applied energy. This understanding was drilled into them when they were not old enough to consume adult beverages. In response, the author offers the following "Silly Little Experiment"; it illustrates why the concept of work is horribly flawed.

The Silly Little Experiment

It may not be obvious to some readers but, the work (force × distance) done to the projectile by the cart's mechanism varies during the Cart Experiment. Logically, it should not since no difference exists from the mechanism's perspective between the two phases. In Phase 1, the cart's mechanism unquestionably does 25 joules of work to the projectile. In Phase 2, the same mechanism used in the same way now seems to do 75 joules of work to the projectile which, by any standard should be impossible. This 75-joule value comes from taking the amount of kinetic energy the projectile has at the end of Phase 2 (100 joules) and subtracting the amount it has at the beginning of that same phase (25 joules). Naturally, physicists can explain this. The problem is that their explanation avoids the mechanics of the situation.

The reason for the varying amount of work (force × distance) is quite simple. Force is a constant throughout the two phases of the experiment but, the distance it acts through is not. It is wholly dependent on the cart's initial velocity. During Phase 1 when the cart begins at rest, the force provided by the cart's mechanism acts through a certain distance. When the same force acts in Phase 2, the cart is no longer at rest, it moves at 5 m/s. This has the effect of artificially increasing the distance force acts through. This is because the projectile accelerates through two distances concurrently. One of these relates to the distance travelled in Phase 1 and it gets added to the distance the cart moves during Phase 2. Interestingly, the time force acts in both phases is the same.

Anyone can understand the mechanics of what happens by doing the following Silly Little Experiment. The only things required are a few friends, a stop watch, a tape measure, and a car. In a safe location with the driver keeping the vehicle at rest, have someone in the backseat hand his cell phone to the passenger in the front. Measure the distance the phone travels and how far the car moves (none during this time); add those together. Also, time the event, either in the car or from a convenient location outside. Next, get the car moving at 1 mile per hour and repeat the exact same action of handing the phone from the back seat to the front. Re-time the event in the same way as before. Once again measure how far the car travels and add that to the distance the phone travels inside the car. This produces the total distance the phone travels with respect to the ground. As should be obvious, the ground serves as the requisite reference point for all measurements in this experiment.

Comparing the data finds that the times are the same throughout within any reasonable margin of error. On the other hand, the distance the phone travels varies wildly and depends utterly on the car's velocity. Since force is the same throughout, the amount of work done to the cell phone changes due to the artificial increase in the distance traveled. If the car were to travel at an even greater speed, the time to do the deed would continue to remain unchanged. In that situation, the total distance the phone travels increases once again, and the work done would increase accordingly. In short, the work done to the phone varies because the distance the phone travels increases artificially. It does in fact travel through two different distances at the same time. This means that force × distance is a useless concept. It varies because the distance force acts through can vary for reasons that have nothing to do with the amount of energy in play.

A question for anyone who disagrees; does it take more biological energy to move the cellphone from the back seat to the front when the car is moving at 60 mph compared to when the car is at rest?

This Silly Little Experiment, all by itself, invalidates the idea of work and by extension the kinetic energy formula. It shows the work done by the cart's mechanism increases artificially simply because the cart is in motion during Phase 2.

CHAPTER 8

The Astronaut Wrench Experiment

In the previous chapter, the Silly Little Experiment illustrates the undeniable bond force has with time, not distance. This chapter confirms that bond in an entirely different manner.

Take the case of an astronaut floating in space who decides to throw a wrench. Newton's Third Law often quoted as, "for every action there is an equal and opposite reaction," tells us that force acts in two directions simultaneously. Basically, when we push on something, that something pushes back with equal force. This is particularly evident in space where the effects of gravity and other factors cannot obscure what occurs.

As the astronaut throws his wrench, both he and the wrench accelerate away from one another. Neither the wrench nor the astronaut experiences force unless the other one does as well. This means that both the astronaut and the wrench accelerate for the exact same amount of *time*; the distance each accelerates through will however be different. And in case anyone has any questions about this, the distance each travels begins at the location where the astronaut was once holding the wrench; they are just measured in different directions from that point.

Astronaut Wrench Experiment

Recall from both the Cannon Analogy and the Explosion Scenario that two objects being pushed apart always result in both objects having the same magnitude of momentum; the objects just travel in different directions. This translates to a situation where the astronaut and wrench receive the same quantity of dynamic energy (mass × speed).

Kinetic energy wise, the only time two objects receive the same amount is when their masses are the same. This was illustrated in the Explosion Scenario in Chapter 5. In the Cannon Analogy, as depicted in Chapter 4, the mass of the two objects were different; the "projectile" had a mass of 1 kg, the "cannon" 4. Using the kinetic energy formula and velocity values from that example, the "projectile" receives 200 joules of energy, but the "cannon" only gets 50.

The best way to understand why the Astronaut Wrench Experiment invalidates the kinetic energy formula is by shifting viewpoints. Consider the situation where we have a compressed spring between the astronaut and the wrench. When it releases its stored energy, the spring causes the wrench to accelerate in one direction and the astronaut in the other. This results in the wrench acquiring mechanical energy in a manner similar to what occurs with a bullet. This analogy continues with the astronaut having the status of a rifle while the spring acts as the gunpowder. If we look at this *exact same event* from a different perspective; the astronaut could be considered a spaceship. In that case, the spring acts as the energy source with the wrench being the propellant. In both circumstances, the spring must release the exact same amount of energy to the "bullet" and to the "spaceship" for it can do nothing else. In short, an energy source cannot create two different numerical effects when nothing changes whatsoever except how it is viewed. So, the astronaut and wrench must wind up with the same amount of mechanical energy.

A Word on Springs

The construction of springs seems to suggest they align with the concept of work. This is because one must apply a force that acts through a distance to fully compress them. The thing is that it also takes time to compress a spring.

The other thing about springs goes to their construction. They are typically fabricated using steel and so, they have some mass. This means springs are in a companion law situation. For example, a spring might vibrate back and forth after the object it accelerates is no longer in contact with the spring. This would occur for objects with too little mass. If the object's mass is exactly right for a particular spring, it does not vibrate. In such cases, the spring releases 100% of its stored energy, most of it to the object with the rest causing any wind and heating effects. In the Astronaut Wrench Experiment, the spring accelerates an object with exactly the right amount of mass (the astronaut) and one with too little mass (the wrench) at the same time.

Force and Time

Believe it or not, physics instructors teach students about a different physics principle that showcases the unique bond force has with time, not distance. It occurs during the basic lessons on electricity. Physics professors teach that a quantity of electrical energy is equal to voltage × current × time. This also represents the usual method by which utilities charge for electrical power. Many small businesses and practically all homes get charged by the kilowatt hour. A "kilo" means some value gets multiplied by 1,000 and a watt is the named unit for the product of voltage and current; an hour is a unit of time.

To produce mechanical force using electricity, it requires applying voltage to certain kinds of electrical equipment. This causes an electrical current to flow. This then translates to voltage × current and results in mechanical force when applied to an electrical motor. In other words, a quantity of electrical energy relates directly to *force* (voltage × current) multiplied by *time*.

The Two Doors

Today, science teachers use perfect or ideal objects in their lessons. Those mythical objects trace their origins back to Descartes and momentum's development. After that, Huygens investigated $m\mathbf{v}^2$ and later, Newton revealed his Laws of Motion at which point, the use of perfect or ideal objects became set in stone.

Once Leibniz introduced Vis Viva and, certainly when Dr Young used the word "energy" in his public lectures, someone should have looked into momentum. This is the first door to walk through whenever a new discovery or better data turns up. In other words, scientists are supposed to review older ideas for signs of weakness whenever a new discovery or better data suggests the need. It is a wise scientist who realizes that knowledge evolves over time; what was true yesterday is not always true today, or will be tomorrow.

As for the second door; it leads to new avenues of exploration including new technologies. Historically and even today; physicists tend to focus more on the second door. For example, newer and better observations of the heavens indicate a problem with their understanding of gravity. To address those concerns, physicists invented "dark energy" and "dark matter". They did not open the first door wide enough to walk through; they just glanced in that direction. And even after searching for years in vain for dark energy and dark matter, they still avoid walking through the first door.

Point Made

There you have it; overwhelming evidence today's physicists were poorly taught by others who had the same misfortune. Once again, the author asks physicists to verify everything in this book. Before doing so, they should examine the Academic Experiment described in Appendix D. It is a case study in how not to confirm new hypotheses or review existing theories.

The rest of this book assumes the reader agrees or is at least willing to concede the possibility that effort (force × time) and dynamic energy (mass × speed) are the proper mathematical expressions for mechanical energy.

CHAPTER 9

Definitions

This chapter only provides the definitions for those physics concepts directly related to the New Hypothesis. Be forewarned, one of the definitions is quite unique; it clarifies an important fundamental tenant in physics that physicists always get wrong.

The Universe

If anyone is going to study the universe, it behooves them to figure out its constituent parts and define each. These break down into four relatively obvious concepts: energy, time, matter, and space. There is one more obvious concept but, many physicists may consider it too controversial and so, the author is not going to name it at this time.

Matter

Matter is the stuff we can touch, smell, taste, and so on; it can be further quantified in many ways. One important one is mass, how much stuff in an object. Another would be volume, how much space does it take up. And yet another would be density, how much stuff within a given volume. There are also other ways including color, natural state at room temperature and so on.

Solid, Liquid, and Gas

Matter exists in three states: solid, liquid, and gas. In solid objects, the atoms and molecules are trapped in place where they vibrate. In a liquid or a gas, the atoms and molecules still vibrate but they can and do move about. So, in a sense, a great deal of similarity exists between liquids and gases. In other words, and not necessarily as a proposal of anything, matter could be said to exist as a solid and a non-solid. Despite this observation, gases and liquids might be better described as two different states of matter while also noting their similarities. For example, both tend to fill any container in which they are placed. Gas fills the entire container whereas, a liquid fills from the bottom on up and not necessarily the entire container.

Mass and Weight

Mass and weight are closely related but they are not the same. Mass is a measure of an object's resistance to changes in motion. Mass is also a measure of how much stuff in an object and it does not vary with location. The greater the mass, the more difficult it is to change an object's speed. A tennis ball has far less mass than a similarly sized solid iron cannon ball. Consequently, a tennis racket has no difficulty colliding with one and nothing but trouble with the other.

Weight, on the other hand, can and does vary depending on an object's location. Despite this difference, if location does not change, mass and weight can crudely be thought to be the same. Unfortunately, mass and weight have the same named units and that is an issue requiring resolution. The ideal unit for weight is the newton, the SI unit for force. The unit for force in the Imperial System of Units is the *poundal*. However, through years of misunderstanding between weight and mass, the word "poundal" has vanished from use. James Clerk Maxwell did however use the term in his 1877 book, "Matter and Motion."

To determine the correct weight of an object, take the value given by a scale and use Newton's Second Law, **F** = *ma*. Weight could be defined as the force an object produces due to the effects of gravity or acceleration. It is equal to the amount of mass an object has multiplied by the effect gravity produces. Remember that on Earth, 1 kg of mass and 1 kg of weight are basically the same. This means the following formulas would not work on the moon or any other celestial body. In terms of newtons, the reader can calculate his or her actual weight using one of the following formulas. They assume the effect of gravity is about 9.8 m/s². Please note that gravity is quantified in terms of acceleration because that is the effect gravity creates.

$$\text{weight in newtons} = \frac{\text{mass in pounds}}{2.2} \times 9.8 \text{ m/s}^2$$

Formula 20, Calculating Weight using pounds

$$\text{weight in newtons} = \text{mass in kg} \times 9.8 \text{ m/s}^2$$

Formula 21, Calculating Weight using kilograms

There may be a few ladies who would rather be known for weighing 110 pounds than for weighing 490 newtons. If that is the case, they could always say that they have a *mass* of only 50 kilograms (110 pounds ÷ 2.2 = 50 kg).

Energy

Energy is basically that which can create change; it is essentially the concept we call cause. Whenever an object moves, it first took energy to change its status from an object at rest to something that is in motion. Once the object is in motion, it has energy because it moves; it is continually changing its position in space. The faster it moves, the greater the changes in distance per second and thus the more energy it will have.

Force

The word "force" is one of those words that has been around for a long time; its roots go back at least 1,500 years to the Latin word "fortia". Force is another word for "energy" when energy tries to or causes a change in motion. Force, like the concept of energy, identifies with the concept of cause. Energy is the more general term while the word "force" is specific to a push or pull as the action of energy.

Power

In the automotive world, manufacturers use the term horsepower as a selling feature. Typically, the more horsepower a sports or muscle car has, the more it will set you back at time of purchase. Horsepower is the rate at which work (force × distance) is done and that equates mathematically to **F × d ÷ t**. This is an improper measure of power that is better defined by the concept of force alone. Power equates to the strength of something. An engine or motor that provides 100 newtons of force is less powerful than one that provides 500.

Mathematically, the strength of anything is equal to the effort it applies divided by time; not surprisingly, this brings us back to the word and concept of force. The physicists of the past got the dividing by time part correct; their mistake was beginning with force × displacement.

$$\text{Power} = \frac{\text{effort}}{\text{time}} = \frac{Ft}{t} = \text{force in newtons}$$

Formula 22 Power

In electricity, a similar situation exists. For example, if a toaster's power rating is 1,000 watts, the amount of electricity it uses for a given period is the product of its power rating and time. So, if you use a 1,000-watt toaster continuously for one hour, you will have used 1 kilowatt-hour worth of electricity; a kilowatt is equal to 1,000 watts and when that is multiplied by an hour you get a kilowatt-hour. Recall that a quantity of electrical energy is related to the product of watts (volts × amperes) and time. Power is simply how strong something is at any given moment.

Time

Time is something everyone deals with and yet, it is one of the most misunderstood concepts in physics and in life. At its very core, time is nothing more or less than *the fact that change occurs*. This is time's basic definition. This may not seem correct or even understandable until one examines time more closely. To understand time's fundamental definition and its nature, it is only necessary to ask how we know that time passes. Clocks tell us but, they are both artificial and arbitrary. The best answer is that we know time passes because things change. If any one of us were to wake up in a hotel room that looked exactly like those that existed a hundred years ago, we might think we went back in time. Once we left the room's confines and saw the world as it is now, we might instead conclude that time had stopped inside that hotel. Change is what defines time; if nothing changes whatsoever, time does not exist.

The importance of time as a concept is two-fold. Outside of the science of physics, time provides a point of agreement. The party starts at 8 pm and we all have access to clocks. We agree to show up by 8 o'clock and have fun. Many people get paid by the hour; eight hours passes, and both the employee and employer agree that $160 needs to change hands.

Within the science of physics itself, time has an entirely different use and purpose. Understanding time in this context begins by noting that different events take different amounts of time. This can be seen in two ways with the most obvious requiring a time piece of some sort. The other method and the one that does not rely on anything artificial is a direct comparison between two different phenomena. For example, in the time it takes the Earth to complete one full orbit about the sun, the moon will make slightly more than 13 trips around the Earth.

Clocks and calendars are useful in physics; they provide a convenient but arbitrary standard from which to compare different phenomena. Using the example from the previous paragraph, the Earth takes more than 31 million seconds to orbit the sun while the moon takes a bit less than 2½ million to complete a single trip around the Earth.

The basic unit of time in physics is the second and it is to that degree a point of agreement. Over the years the definition for 1 second has evolved into its current standard. It was once defined as $1/_{86,400}$ of a day (there are 24 hours in a day × 60 minutes in an hour × 60 seconds in a minute = 86,400 seconds in a day). For most of us, this definition remains quite adequate. If the truth be known, its modern definition has little meaning to most of us and so, it is not given here. That is not to say the modern definition is unnecessary. The Earth's rotation is not fixed; it has changed over the centuries and will continue to change.

Clocks are an interesting study in and of themselves. Unfortunately, some physicists think time and clocks are the same. They do not seem to realize that different clocks react differently under identical conditions. Clocks are made from the materials available in the universe. They are therefore subject to the various laws of nature that apply to those materials and how they are used. This means some clocks will speed up in some circumstances while others slow down. Moreover, some clocks work in some locations but not in others. There is no perfect or ideal mechanism for a clock although today's atomic clocks do come close.

A pendulum-based clock, for example, will not work at all floating in space. If placed on the moon, it shows time moving slower than it does here on Earth by a significant amount. Place that same pendulum-based clock on a rocky planet the size of Jupiter and time appears to speed up dramatically. In these other worldly situations, time does not actually vary as indicated by the clock. By comparison, an inexpensive wristwatch that uses a mainspring would in those three non-earthly situations be far more accurate. It would also remain in sync with its twin here on Earth for far longer.

Physicists sometimes debate or talk about why the arrow of time only seems to move in one direction. They may point to the fact that you can break an egg and then remark that the reverse cannot happen easily or at all. Some physicists point out that nothing in the known laws of physics prevents time from going the other way. Evidently, they like to muddy things up because physicists do not seem to understand that time only moves as it does because time records the fact that things change. Effects do not make changes; causes do and so, time only moves the way it does because that is the way things are. Sometimes the simple and obvious answer is the correct one.

Despite what Einstein asserted, speed is not a factor in how time progresses for a clock or anything else; acceleration however is. If two clocks are synchronized, one traveling at nearly the speed of light with another identical one at rest, they will remain in sync. And they will continue to do so until one of them accelerates or experiences the effects of gravity.

Every experiment that seems to confirm the effect speed has with time was done while wearing a specific type of blinder, aka confirming a "proven" fact. The various conditions that affect clocks were never fully considered. A world of difference exists between looking for a specific result in an experiment and analyzing that same experiment while not wearing blinders. In the first case, achieving the expected result whether real or imagined is as far as one needs to go. This is not proper; a scientist is supposed to look for and find any factors that may play a role in the experimental results; think companion laws. Physicists only seem to do this whenever someone makes a claim against conventional wisdom.

Both acceleration and the effects of gravity can affect time pieces because time pieces are made of matter. Mass is the property of matter that resists changes in motion. Hence when clocks or other objects including atoms and molecules are subject to acceleration, parts will begin weighing more. This is especially true in extreme cases of acceleration or gravity. The most important thing to remember about time pieces is that they are constructed using the materials of the universe. Therefore, clocks are subject to the various laws of nature including the fact of companion laws.

Space

Space is merely *the fact that a separation exists*. Object A is separate from B, and it is space that is between them. The thing about space is that it comes in three distinct dimensions. One-dimensional space is simply a straight line between two points. Any location on that line can be mapped by comparing its position to a different point on the line.

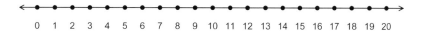

One Dimensional Space

Two-dimensional space is known as a plane, a flat surface akin to the top of a table. It is graphed with two one-dimensional lines where one is drawn at an angle of 90 degrees to the other. Any location on that plane or surface can be identified by comparing its position to a point on *each* of the two defining lines.

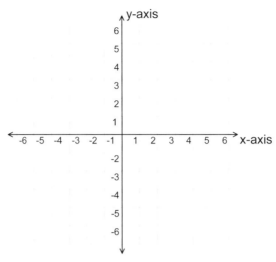

Two-Dimensional Space

Three-dimensional space defines a volume; it has height, width, and depth. It is typically shown on a two-dimensional surface (a piece of paper or monitor) with three one-dimensional lines each of which is at a purported angle of 90 degrees to each other. Examine any one of the eight corners in most rooms and one becomes a witness to a more realistic understanding of the 90-degree angle between the three lines. Any location can be mapped by comparing its position to points on the three lines.

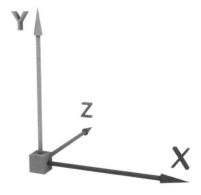

Three-Dimensional Space

A fourth line, drawn once again at 90 degrees to the previous ones, cannot happen. There are no other spatial dimensions beyond the three. Time is not a dimension of space; it is its own unique and basic universal component. Time is however an intimate *companion* to space. In describing an object, it is necessary to provide its location and when it is there. This is because things made of matter motivated by energy can and do move through space over time.

Space contains all sorts of phenomena but, space itself is none of those things. It is just the fact of separation between two points. Space does not necessarily need to house other things such as energy and matter but, it always seems to do so. This is merely a coincidence, an incidental fact whose importance should not be overestimated.

Beyond the furthest extents of the universe, no space exists. Space can only exist between two points of energy or matter. So, as the light from the stars emanates outward beyond the confines of the universe, space is constantly created quite literally from nothing. Conversely, space vanishes between two objects or points as they approach one another and so, space is fluid but only to that extent. There is no law of conservation of space in the usual sense. Moreover, space cannot be bent, stretched, or compressed; it can however appear to do so.

In the final analysis, space is the playing field for the physical universe, and it comes in three distinct dimensions. It is not made up of energy or matter but houses these things. Space is its own unique fundamental and a necessary component of the universe along with energy, matter, and time. Just for fun, take a moment and try to envision a universe that has no space; it cannot be done, not really.

Space is quantified by means of an arbitrary system. The easiest system is the International System of Units; it employs the *meter* as the basic unit. In the US, it is possible to say the basic unit of measure is the inch, the foot, the yard, or the mile. With the International System, the differences between different units boils down to powers of 10. A centimeter is $1/100$ or $1/(10 \times 10)$ of a meter. A kilometer is 1,000 meters (10 × 10 × 10 meters). In the US, a foot is 12 inches, a yard is 3 feet, and a mile is 1,760 yards. One of these systems is far more logical and easier to learn and use than the other.

The idea of space-time is built on two different fundamentals. This Einsteinian invention is the union of the fact things are separate from one another and that the separation between objects changes over time. Momentum is also built on two basic ideas; one is dynamic energy, and the other is the fact space exists in three dimensions which, by necessity, includes the idea of direction. This explains why momentum, as an improper physics principle, was so useful in rocketry and appeared valid to physicists. It also helps explain why the flaws within Newton's Laws of Motion went undetected for as long as they did.

Something to Ponder

If *time* is the fact things change, *space* is the fact of separation, and *energy* is the concept of cause; what is matter? Could *matter* be energy that has, to one degree or another, become effect?

Beyond understanding the key words relevant to basic physics, it is necessary to understand the various tools relating to the subject. Mathematics is unquestionably an important one, as is using the scientific method. Understanding the anatomy of illusions is another and it is the subject of the next chapter.

CHAPTER 10

Illusion Theory

Physicists have long sought to unlock the nature of nature never realizing the universe is a study in illusion. An *illusion* is an observation or data that suggests a cause or a reason that is not the correct one. For example, take the classic magic trick of pulling a rabbit out of an empty hat. If the truth be known, the hat was never empty, or the rabbit came from somewhere else.

Physicists would fare far better as scientists if they considered the universe's creator, whether it be a supreme being or not, the greatest illusionist that ever was. And part of what makes this grand architect so accomplished is that some of his constructs are not illusions at all. Sometimes the obvious answer is the right one. At other times it is not thereby adding a unique and diabolic complication.

Interestingly, the questions that unlock natural illusions also work in other areas of life. This includes exposing the fraudulent narratives that are part and parcel of everyday living. Unfortunately, it is far too easy to fool people. Some are more susceptible than others and yet, no one is immune.

To unlock an illusion, one need only ask the right questions and of course, get the appropriate answers. One interesting fact regarding the various question areas is that they stand firmly on their own but, they can also intersect. One set of questions might unlock an important fact that can also be discovered using questions of an entirely different area. This is not a failing in their classification but a demonstration of their inherent power and efficacy.

1. The Observer

The position of an observer is often quite important to an illusion. Some of the illusions stage magicians create are only possible because the audience observes them from a single direction. This means the audience only sees those things the magician allows. Whenever audience members surround a magician during a televised special, these witnesses are in on the illusion. They see how the trick is done but, the television audience only sees what the magician wishes.

Standing here on Earth, we do not feel the Earth spinning and so, we have the illusion of a sun rising in the east and setting in the west. This illusion was unlocked by observing the heavens and later confirmed from space. This is the best place to expose the illusion of a rising and setting sun. If you sit at the right location away from the Earth long enough, you see the planet rotate. Stated slightly differently, the sun does not move across the sky, it remains in place while the earth rotates on its axis.

The one thing to remember when selecting the optimum viewpoint is that no single viewpoint is always perfect. In the sun rising and setting example, that illusion vanishes when the sun is seen as stationary. In truth, the sun moves; it maintains an orbit within the Milky Way Galaxy. We can only consider the sun stationary under certain circumstances.

Part of selecting or using the optimum viewpoint is the awareness that blinders do exist. These are formed by one's experiences and education. They can hide the true nature of a situation in a way nothing else can. This is the number one reason a science stultifies and why scientists summarily dismiss new ideas. Authorities think they know all about something and so, anything new must be wrong. It never occurs to them that knowledge evolves over time; what was true yesterday is not always true today, or will be tomorrow.

The concept described by the word "superficial" also falls into this first question area. It refers to looking at something without realizing that some observations are the result of two or more actions occurring concurrently. Think back to the Silly Little Experiment at the end of Chapter 7. It shows how and why force × distance is a horribly flawed idea. It did that by doing nothing more than moving a cellphone inside a car. When the car is not moving, the phone travels about three feet. When the car is in motion, the phone moves a great deal further because it travels through two different distances at the same time. One of these is the distance within the car itself; the other is the fact the car also moves. This adds to the total distance the phone moves with respect to the ground. The beauty of the Silly

Little Experiment is that it is painfully obvious the cell phone moves through two different distances at the same time.

The first question one should ask of him or herself is, "Am I viewing the situation from enough viewpoints?" The second is, "Is one or more of these viewpoints the optimum one? In short, any question that seeks to locate the optimum viewpoint and or uncovers any pedestrian or superficial observations is a question worth asking. And by viewpoint, it is meant the location, whether literally or figuratively, from which something is viewed. This also includes the distance from which someone observes. For example, some works of art when observed too closely appear to be nothing more than a collection of colored dots. Sometimes, a mural or other picture is a collection of miniature photographs acting as pixels. The artist's intent only becomes knowable when observing his or her painting from the right distance.

2. Relative Importance

Not all facts have the same value; sometimes one fact is extremely important. In other situations, the same fact has no value. Magicians often emphasize actions which play no role in the outcome of a trick, aka misdirection. This is often given as the most important tool a magician uses when it is merely one of them. Misdirection is the art of changing the importance of actions.

This question area is particularly relevant in physics as it relates to the physicist's favorite and sometimes only tool, aka mathematics. When one knows the numerical values of two or more events, a comparison becomes possible. And part of making measurements is the necessity of knowing how precise or imprecise those measurements are.

Part of relative importance is knowing a concept that evidently goes untaught. It could be called the *incidental fact*. It is defined as something that can be seen, measured, and verified but has no importance or value. An incidental fact is just something that happens to be, aka a coincidence.

Take the case of two strangers about to board a bus. It is easy to verify both have exactly $5. This fact becomes an incidental one should only one of them be allowed to ride. If the bus fare is $3 and must take the form of exact change, the poor soul with a $5 bill is out of luck; the other fellow who has five singles gets to ride. In this situation, the only important fact is who has three one-dollar bills with him. The fact they both have exactly $5 is a detail that is not important in this situation.

Sometimes a book, a lecturer, or other source of information passes on a demonstrably false "fact." If it has no importance in terms of a conclusion or specific purpose, it does not lessen the value of the other statements from that source. Such things are merely minor errors. For example, if a history textbook states that General Ying was 5 years older than General Yang, that fact may very well be wrong but, it does not matter. The important thing is who won the battles between them. The relative age of the Generals has no bearing in such situations. That fact could be omitted without changing the described historical event and its importance to the two nations involved. In other words, do not worry about the insignificant errors people sometimes make. Errors only become important when they affect a conclusion.

One should determine the relative value of the facts. It is imperative to know what facts are useless, are coincidences, and which ones are important.

3. Time and Distance

Time and distance have a particular importance when it comes to illusions in physics. These two concepts are related; one cannot exist without the other and yet, each stands on its own within its own realm. In short, they are often intimate companions.

Sometimes when a magician performs, the "magic" occurs long before the audience suspects. For example, the trick called Metamorphosis involves an assistant and the magician. In it they appear to switch places in the blink of an eye. Factually, half the trick is done long before the audience suspects; the other half occurs when the audience assumes the trick is over but before the final reveal.

Motion occurs when something travels through a distance over time. It is therefore vital to note both *time* and *distance* in any physics experiment where motion is a factor. Moreover, nothing can change instantly; time always passes even if it occurs extremely briefly, and that fact must be known. Furthermore, in any observation where motion occurs, changes in distance are generally more noticeable than time and so, always keep *time* in mind. Please note that because the author understood the link between time and distance, he was able to see the blunder Du Châtelet made and find the related flaw within the Academic Experiment (see Appendix D).

As a rule, sequences have a great deal of importance. In physics, it is best to unlock the simpler aspects of nature before tackling the more complex. Moreover, the wrong sequence is to suppose something is true and then go about writing a Law or working out the math that validates the supposition. Descartes violated this when he developed momentum. In science, the proper action is to follow the sequence defined in Appendix C, "The Scientific Method."

Sometimes, the concept of distance is far more important than time; for example, when it comes to magnetism and gravity. Interestingly, at smaller scales, magnetism is the emperor that rules with an iron fist. At larger scales, gravity is the tyrant whose word dictates reality.

The questions with respect to time one must ask are quite logical. Has time been noted as is appropriate? Has a proper sequence been violated? In physics particularly, one must understand the nature of time (See Chapter 9). Distance, along with time, must be noted in any situation involving motion of any kind.

4. Assumptions

Magicians orchestrate their illusions in such a way as to cause the audience to form certain reasonable assumptions. For example, some of their equipment appears too small to conceal a human being even upon close-up visual inspection. One trick magicians use involves using walls made of rubber; rubber stretches allowing a small space to expand when needed. Another trick includes painting parts of a prop so they are the same color as the rear curtain. These and other similar tricks make it easier for the audience to assume a box is too small to contain a human being.

Sometimes, a physicist wishing to advance the subject will make a supposition. This should be avoided whenever possible. However, if someone feels the need, that supposition should be made with extreme care and knowingly. If it is in error and it remains unrecognized as an assumption, a blinder may be forged so well that it becomes extremely difficult to expose. For example, since the time of Descartes, physicists have assumed any discussions related to how fast something moves must also include a directional component. Consequently, physicists think almost exclusively in terms of velocity, rarely about speed. This prevented Leibniz, Newton, and Du Châtelet from considering the merits of the product of mass and speed as an option for Vis Viva.

A subject that rests on too many unverified assumptions is bound to go awry. Descartes began with a supposition that no one took the time to verify until recently. When it was finally checked and found wanting, that action led to an entirely different view of the universe, with a surprise or two still to come.

Physicists might wish to periodically list those assumptions they were taught and the ones they make themselves. Test the ones that can be verified and then re-evaluate the rest. Sadly, most of us make far more assumptions than we would care to admit. One of the first ones everyone makes is we assume that our physics professors teach proven facts. For centuries, teachers told students momentum was always conserved and every one of us incorrectly assumed it was true.

Assumptions sometimes take on an insidious form known as the false premise. Whenever a scientist, an investigator, or anyone else begins with one of these, the truth becomes elusive. A false premise is akin to putting on a blinder that obscures or colors everything learned thereafter. If the reader examined and understood Appendix D, "The Academic Experiment," they can say they were a witness to the deleterious effects of a false premise. And in case anyone does not know what that false premise is, it is that if an experiment confirms $\boldsymbol{Fd} = \tfrac{1}{2}m\boldsymbol{v}^2$, it confirms that kinetic energy is the proper measure for mechanical energy.

Ironically, it is not a bad policy to assume that assumptions exist whenever one begins investigating anything.

5. Lies

Not everything the magician says is true. He might state there is nothing up his sleeves when there is. For example, magicians sometimes carry a flesh-colored false thumb in which they can hide small items. Even from a short distance, this thumb placed over a real thumb is virtually impossible to see. It is especially difficult if it is a professional thumb used by a practiced magician.

In other areas of life, lies are ubiquitous and frequent. This is especially true during political campaigns and on the internet. For example, a description of a video on the internet might imply or even make a bold claim. However, after viewing it, the obvious conclusion is the description was either a bold-faced lie or an exaggeration.

When doing an experiment, sometimes the equipment may not read correctly. The value shown on a meter might say 10 volts when the actual value is something entirely different. Beyond equipment failures or errors in reading displays there are other sources of lies that can be extremely problematic. For example, Descartes's momentum is not a true representation of reality. Even the use of ideal or perfect objects when teaching physics is a lie. And with such and other lies being the norm in physics, the search for the true nature of nature suffers.

Uncovering lies can take some leg work. It begins by verifying the integrity of any equipment and knowing, not assuming, the currently accepted laws of physics are indeed valid. One part of this key is not to fall for the same tricks many news and intelligence organizations do; if two or more different reliable sources tell the same lie, many news outlets assume the lie is a fact. Two different sources of information may very well be honest but, if each got their data from the same misinformed source, a "confirmed fact" becomes a lie masquerading as the truth.

Lies that are repeated often enough can go through a transformation; they can become one of those things everyone knows is true. The Law of Conservation of Momentum is just one example; it was deemed true by every science teacher, physicist, and textbook for more than three centuries. It was originally assumed true by the likes of such notable individuals as Leibniz, Newton, and Einstein but alas, it is not.

Due to how poorly logic is taught, physicists do not seem to know that when one theory conflicts with another that something is horribly wrong. In such cases, the problem might be with either theory or both. The two accepted theories being referred to at this moment are the theories of Relativity and Quantum Mechanics. These are well known within the physics community as being incompatible with one another under certain conditions. Despite this, physicists generally consider both theories to be valid. Factually, at least one of them is wrong in some significant way. In general, if two facts or theories conflict, it means that at least one is a lie or contains a lie. The only caveat to this is the fact that on rare occasions, an apparent paradox is not one at all. In such cases, it is still necessary to investigate and learn why the paradox seems to exist. Humorously, an apparent paradox that is not actually a paradox is itself a lie.

6. Missing Data

Magicians invariably fail to reveal everything they do. If the audience knew everything the magician knew, no illusion would be possible. The only way a magician sees an illusion himself is by assuming the audience's viewpoint and forgetting certain facts. For some tricks, magicians use identical twins. The audience sees a beautiful girl teleport from one location to another in the blink of an eye. The audience thinks it is the same girl when the first sister hid as the other jumped into view from a different hiding place. The missing data is the use of identical twins, a very usual thing for magicians.

When someone is missing key bits of information, it is extremely easy to make a mistake. Du Châtelet failed to record half the data the s'Gravesande Experiment produces and it caused her to form an incorrect conclusion. To this we can add the fact no one ever considered the product of mass and speed as an option to describe mechanical energy, a missing option. If any 17th century scientist had reviewed Du Châtelet's work, noticed time was missing, and included mass × speed as an option, that scientist would have rendered this book unnecessary.

Spotting missing data requires asking questions. These tend to depend on the situation. Obviously, it is easier to see what is there and so, there must always be a concerted effort to look for things that are not there but should be. Chances are that if everything makes perfect sense and all the known facts align perfectly with one another, you may have all the data you need.

The Scientific Method

The winning hand in physics demands using the Scientific Method. It also requires asking the right questions so as not to fall for any illusions.

11

The Dynamic Effort Hypothesis

This chapter adds to and clarifies the New Hypothesis first mentioned in Chapter 6. It begins by changing its name to the *Dynamic Effort Hypothesis* (DEH). This is the official hypothesis the physics community needs to verify. This chapter ends by proposing certain minor but necessary corrections to Newton's Laws of Motion, sorry Isaac.

The Basic Equation

$$Ft = ms$$

Equation 16, the Basic Dynamic Effort Equation

The Basic Equation applies to those situations where an object begins at rest and only accounts for half the action. Stated a bit differently, when used to describe the actions of a cannon ball being fired from a cannon, this equation applies either to the projectile or the cannon. If we wish to examine both the cannon and a projectile at the same time, we need to know force acts in opposite directions simultaneously. This means we could use the basic equation twice, once for the projectile and again for the cannon. We could also expand the basic equation to create a modified version. Either of these approaches would work and they would give an idea of how much energy a source provides. A full accounting requires one more thing and it is described a bit later.

$$2Ft = m_1 s_1 + m_2 s_2$$

Equation 23, Expanded Dynamic Effort Equation

The Expanded Equation is based on the following graphic showing what occurs in many situations when force acts.

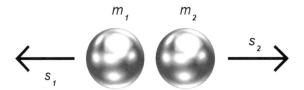

Aftermath of Force Acting

Please note that if the mass of the two objects were the same, both would accelerate to the same speed. Mathematically, Equation 23 could then be simplified or reduced to *Ft = ms*. This should not be done because if it were, it would obscure the fact force acts in two directions simultaneously. More than that, it would produce a numerical value that is half of what it actually is. So, be careful applying the rules of mathematics to physics; in certain situations, some of those rules should not be used.

The Basic Equation allows us to define two concepts; one is for *effort* and the other is for *dynamic energy*. The word "effort" would be used in those situations where the word "work" had been used previously and fills the same need. The phrase "dynamic energy" replaces "kinetic energy." The formulas for effort (*EF*) and dynamic energy (*DE*) are as follows. Both variables use two letters and are as recommended by the author.

$$EF = F \times t$$

Formula 24, Effort

$$DE = m \times s$$

Formula 25, Dynamic Energy

The equations listed so far assume the objects begin at rest. If the objects are already in motion, we need to use a slightly different equation. In its simplest form, it uses the Greek letter delta (Δ) which, has the meaning of change in.

$$Ft = \Delta ms$$
Equation 17, from Chapter 6

Equation 17 requires clarification; the use of the change in symbol is not the same as it was for kinetic energy. It does not use the concepts of initial and final; instead, it is about faster and slower. For the record, it calculates the amount of effort that causes a change in an object's speed.

$$Ft = m_1 s_{faster} - m_1 s_{slower}$$
Equation 26, Clarified Equation

Equation 26 never produces a negative or less than zero result. This is because it always subtracts the smaller quantity of dynamic energy from the larger amount. Remember, energy is that which causes or creates effects. If something changes, energy played a role. If there is less than no energy, what happens then?

Equations 17 and 26 only look at one side of an event. In Phase 2 of the Cart Experiment, both the cart and projectile move. The projectile represents one event and so, if we wish to examine the cart, we need to use Equation 26 for a second time.

As a quick and related aside; in any discussion of speed or mechanical energy, it is necessary to specify a reference point or, it must be obvious. Both speed and mechanical energy are relative quantities; their values depend on something else; think companion law.

Equation 27 below represents something physicists call the "Work Energy Theorem." It contains the elements previously discussed plus one additional factor. That factor takes the form of cos Θ. This version exists to encompass all situations. This is because in certain circumstances the amount of force that causes a change in motion or position can be different from the amount applied.

$$Fd\cos\theta = \Delta\frac{mv^2}{2}$$

Or

$$Fd\cos\theta = \frac{mv^2_{inital}}{2} - \frac{mv^2_{final}}{2}$$

Equation 27, Work Energy Theorem

where,

$v_{initial}$ is the initial velocity

v_{final} is the final velocity

cos is the abbreviation for the word cosine (explained below)

Ө is the Greek letter theta (also explained below)

F is the applied force

d is the displacement force acts through

A cosine is the ratio of two sides of a specific triangle type; it is always a number between 0 and 1. The word "cosine" is from a form of mathematics called "trigonometry" (trig). Trig is the math that deals with the numerical relationships of triangles that have one angle that is exactly 90 degrees, aka a triangle with a right angle. Trig adds up to a more inclusive and complicated method of describing directions.

The symbol "Ө" is from the Greek alphabet and is called theta; it refers to the angle between the applied force and the change in motion experienced by the object. The thinking behind the use of co-sine theta (cos Ө) is to calculate the amount of force that causes a change in position or velocity. It is possible to apply a 100-newton force which results in a force of only 50 newtons causing the change in motion.

Interestingly, the purpose of the cos Θ factor is not obvious given its physical location in the equation. It exists to modify the amount force but physicists chose to place it after both force and displacement. Given the rules of mathematics, this placement changes nothing with respect to the resulting numerical values but, it could confuse a few students.

Take the case of a fellow pushing a box as shown in the following graphic. The force this fellow exerts is partly downward and partly in the only direction the box can move. This means that only a certain percentage of that force is pushing the box in the desired direction. The use of the cos Θ factor calculates that amount.

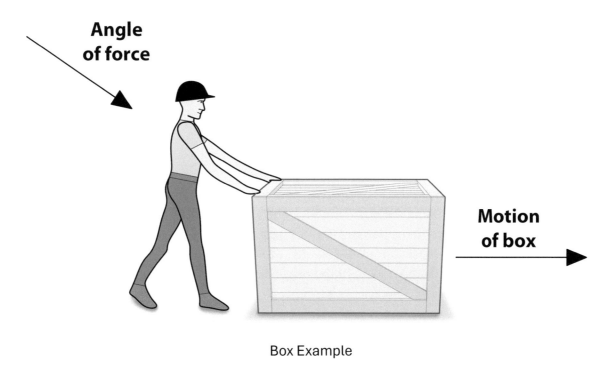

Box Example

In this situation, a portion of the force acts in line with the motion of the box. The rest of the force pushes the box into the floor. In practical terms, the force applied by this fellow is both helping and hurting his efforts to move it. Some of his energy is wasted since it is pushing the box into the ground and not directed towards his goal. That wasted energy is making it tougher to move the box; it causes more friction between the box and the ground. If this person knew a little physics, he would push, not partially downward, but just in the direction he wanted the box to move. This would make his life easier for two obvious reasons. The first is a bit less friction between the box and the ground and, all his energy goes to moving the box in the desired direction. Humorously, he could also use a wheeled cart.

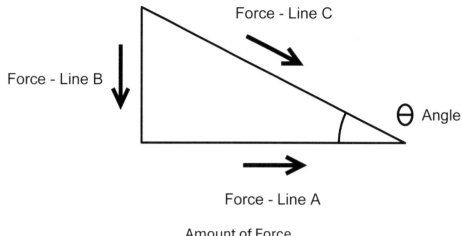

Amount of Force

In the "Amount of Force" diagram, we see three force lines, "A," "B," and "C." The length of the lines represent the relative amount of force in newtons that acts in the Box Example. Line C represents the force applied by the fellow pushing the box. That force can be thought of as acting in two different directions. One is to move the box in the desired direction as shown by Line A. Line B, acts to pin the box to the floor. To calculate the amount of force acting in the direction of Line A, one need only multiply the amount of force of Line C by the cosine of the angle θ. The numerical value of the cosine can be found using a scientific calculator.

Calculating Line B must be done a bit differently. Its value cannot be found by subtracting Line A from Line C or using the cosine of theta. For those who remember their high school math, the following equation might seem familiar and suggests one means of determining how much force pins the box to the ground, $a^2 + b^2 = c^2$. If this equation appears foreign, think unflattering thoughts about your high school math teachers; obviously, they did a poor job.

At this point, it might be a good idea to add just a little bit more to the reader's understanding of friction, thermal energy, and the DEH. The force acting along Line A does not result in acceleration other than when the fellow first starts pushing. Thereafter and during the entire time he pushes the box with a constant force, two things occur concurrently. First, a quantity of energy defined by the applied force acts along Line A. That force tries to accelerate the box in the desired direction. While this is happening, an identical quantity of energy transforms into thermal energy due to the interaction between the box and ground. The surface molecules of the box exert a force on the surface molecules of the ground and vice versa. Believe it or not, this then results in the box moving as described by Newton's First Law of Motion; an object in motion remains in motion unless acted on by a force. The force that might otherwise accelerate the box immediately gets removed and converted into thermal energy. Of course, the moment the pushing stops, the box stops moving in short order.

Understanding the Dynamic Effort Hypothesis requires including an important companion law. It is the thing referred to at the beginning of this chapter that allows for a full accounting of energy. Because introductory physics courses tend to omit this factor, it allowed momentum's flaws to go unnoticed for centuries. Physics professors do however mention that factor in later semesters when they teach the Laws of Thermodynamics. Those laws basically state that whenever an interaction exists between two objects, some quantity of the applied energy transforms into some other form. This energy "loss" should be taught when first teaching about force. Please note that the word "loss" when enclosed in quotation marks does not mean energy vanishes; it means energy transforms into an unwanted form. This definition for an energy "loss" applies to the rest of this book.

While the DEH recognizes certain aspects of the Laws of Thermodynamics; it does not necessarily agree that all of them are valid as written and or understood. To make things easy for non-physicists, those laws and ideas are not part of this book.

The Dynamic Effort Hypothesis understands and includes the fact that nothing is exactly 100% efficient. Sometimes the amount of the energy "loss" is insignificant; at other times it is a rather large amount and so, its existence should always be considered. Any force (energy) that does not result in the desired effect is an energy "loss." Such situations need to be recognized as part of any mathematical description. One must add an additional variable to the various mathematical versions of the dynamic effort hypothesis. This variable takes the form of a single letter "E" with a descriptive subscript of some sort such as "H" for heat as in E_H for example.

In the Box Example, a great deal of the fellow's efforts result in a thermal increase along the ground as well as along the bottom of the box. The rest of his efforts result in moving air molecules about, aka wind that could be represented by the variable E_{wind}. The following equations shows the DEH applied to this Box Example once the box is in motion and moving at a steady rate. When force first acts, the box accelerates and the three energy "losses" increase from their initial zero value to their final amounts.

The variable "F_C" is the force exerted by the fellow and "F_A" is the force acting in the direction the box moves. The variable "t" is the time that force acts, "E_{Hb}" is the thermal energy added to the box, and "E_{Hg}" is the thermal increase that occurs along the ground while the variable E_{wind} represents the energy air molecules acquire.

$$F_A = \cos\theta \times F_C$$

Calculating the force acting in the direction of the box

The values for the three E variables is not fixed. They are entirely dependent on how long the fellow pushes the box. Stated differently, the amount of thermal energy the box receives is dependent on how long the box gets pushed.

$$F_A \times t = E_{Hb} + E_{Hg} + E_{wind}$$

DEH and the Box Example

Please keep in mind that it is not a question of memorizing the various equations. One need only understand the Basic Dynamic Effort Equation and add the necessary embellishments for the situation to hand; think of these as companion laws if you wish. For example, the Δ function applies when the object is already in motion. In such situations, we need only remember to subtract the smaller quantity of energy from the larger amount. In all cases, we use the amount of force that causes a change in motion or position; we only use the cos Θ factor when necessary.

If we need to know the total amount of energy applied, we can use the expanded equation or, better yet, just double up on the basic equation using it once for each side. And finally, there is the use of the "E" variable with its identifying subscript describing the energy "losses." It can be a value that is quite small or rather large and so, it must always be considered and thus looked for in all situations.

Testable Prediction

Every hypothesis must make a prediction, or it isn't science. Ideally, this should be something unique enough to leave little doubt as to the hypothesis's validity. This may not always be possible due to the fact of companion laws but in this case it is. The Dynamic Effort Hypothesis states that an object's dynamic energy increases *linearly* with changes in speed. This prediction is at odds with the one made by the kinetic energy formula. This makes it easy to discover which formulation is indeed valid.

Units

The basic unit for work and kinetic energy was the joule but, that unit cannot be used by the Dynamic Effort Hypothesis. The joule relates to the *displacement* force acts through and it can vary wildly; the DEH relates to the *time* force acts and so, it requires its own unit. This basic unit of energy is defined by a force of one newton acting for one second; it can also be defined by a kilogram of mass traveling at one meter per second. Please note that a time unit is part and parcel of both units.

The word "bob" is hereby proposed, temporarily or otherwise, for the basic unit for energy. It comes from the fact that ships and buoys "bob" up and down in the ocean indicating energy at work (no pun or reference to a force acting through a displacement intended).

The correct unit equation for energy is shown in Equation 28. A unit equation contains no numbers; it merely shows the individual units that make up a more complex one and how they relate to one another. As all physicists know, it can be a useful tool when analyzing new and existing physics formulas.

$$\text{Energy in bobs} = \text{newton} \times \text{seconds} = \text{kilograms} \times \frac{\text{meters}}{\text{second}}$$

Equation 28, Unit Equation for Dynamic Energy

Once the DEH officially transitions from a hypothesis to an accepted theory, it becomes shockingly easy to invalidate Einstein's most famous equation, $E = MC^2$. This is because both the kinetic energy formula and Einstein's equation have the same unit equation; they also both have as their unit the joule. Their unit equation is different from the one defined by the DEH thereby invalidating both.

$$\text{Energy in joules} = \text{newton} \times \text{meters} = \text{kilograms} \times \frac{\text{meters}}{\text{second}} \times \frac{\text{meters}}{\text{second}}$$

Equation 29, Incorrect Unit Equation for Energy

Testing

There you have it, the hypothesis that upon confirmation should replace the formulas for work and kinetic energy. If the Basic Dynamic Effort Equation is tested within its simplest scope of use, the other versions are essentially also verified. The need for the *E* variable has already been confirmed and explained; physicists already understand it and its value.

The Types of Dynamic Energy

Dynamic energy exists as a mass moving in a straight line. Examples of this include a cannon ball fired from a cannon, the Voyager space probes hurtling through space, and anything similar including cars driving down the road. The second type of dynamic energy deals with the vibrational motion of masses manifesting as thermal (heat) and acoustic energy (sound).

The DEH looks to rotating or orbiting bodies a bit differently. Take the case of an orbiting object such as the ISS (International Space Station). Its path around our planet is nearly circular. That motion is the result of two *different* conditions occurring concurrently. One of these is that the space station has a quantity of dynamic energy; it moves in a straight line. The other condition is due to the gravitational effect between it and the Earth. Remove either condition and the ISS's motion changes; it will no longer orbit the Earth.

In practical applications, as in sending rockets into space, it is sometimes best to consider all three dimensions at once. In many cases, two dimensions are all that are required. In terms of trying to unlock the true nature of nature, it is best to break down an object's motion along each dimension or condition separately. In other words, the DEH understands the practical view of describing circular or rotating motion as a phenomenon on to itself. It is particularly important to NASA and others to think of the universe as a three-dimensional structure. However, physicists seeking to unmask the fundamental laws of nature should not think in terms of composite laws. Engineers can use the math describing circular motion as an individual phenomenon all they want. Unless a physicist is engaged in such practicalities, he or she should stick with the fact circular and other complex motion is due to two or more phenomena acting concurrently.

A *composite* law is the union of two or more laws. Momentum is a composite law and it worked for rocket scientists but, it failed miserably in other situations. Space-time is a relatively newly formed composite law that obscures the true nature of the universe. It may have some practical value, but it is not pure physics in the same way as effort or dynamic energy are. Objects inherently move along a one-dimensional line and when they do not, other factors are in play.

Blinders

The only thing that can get in the way of the DEH becoming an accepted theory are blinders. If physicists cannot see that momentum is not the conserved quantity they were taught, they will be unable and unwilling to see the various other revelations including the double standard discussed in Chapter 7. To quote an old saw, "There are none so blind as those who will not see."

Some readers may have noticed that the Dynamic Effort Hypothesis does not preclude energy from vanishing. This may prove to be an even greater hurdle than a simple correction to the mathematical form for mechanical energy. This is because the Law of Conservation of Energy is a major part of the blinders foisted upon physics students. It may be even more important to the physics community than the concepts of momentum, work, and kinetic energy. In terms of sequence, the first thing to do is to verify the proper form for energy; is it defined by $½mv^2$ or *ms*? Thereafter, physicists should determine if energy can be created and or made to vanish. No one should assume energy cannot be created or destroyed and move forward from there; that is doing things in the wrong sequence.

Updating Newton's Laws of Motion

Chapters 3 and 4 show that the prediction Newton's Three Laws makes is not true. When we compare those chapters with the axiomatic nature of Newton's Laws, the result is an apparent paradox. However, that paradox quickly dissolves upon realizing words like "force," "action", and "reaction" relate directly to the concept of energy. Remember, energy was unknown to Newton, and he was, at best, suspicious of its forerunner, Vis Viva. Add an extremely pedestrian understanding of heat way back then and no one can blame Newton for writing his Three Laws in the manner he did. He should be celebrated for getting so much of them correct. The first change to Newton's Laws of Motion begins by simply calling them "Newton's Axioms" or if you wish, "Newton's Three Axioms." The rest of the proposed changes are as follows.

Newton's First Axiom

"Every object perseveres in its state of rest, or of uniform motion in a straight line, unless it is compelled to change that state."

This law only requires a minor re-write while reminding physicists that Newton considered gravity to be a force of attraction. Gravity can on occasion be thought of as a force but, in the grand scheme of things, it is not, hence the phrase "compelled to change that state." It is also a good idea to mention that the dynamic energy of an object can change when the reference point for speed changes. Consider the car-speedometer example from Chapter 7. The vehicle moving at 84 mph has two different quantities of dynamic energy because it travels at two different speeds. It moves at 84 mph with respect to the road and at 20 with respect to a different vehicle moving at 64 mph. This does not alter Newton's First Axiom but, it illustrates the importance of noting a reference point. Reference points amount to a companion law when speed or dynamic energy are the topic.

Factually, nothing can move at a single speed with respect to all other objects in the universe. This includes light and thus we have an interesting reason to invalidate Einstein's basic postulate. He stated the speed of light (C) in a vacuum is always 299,792,458 meters per second regardless of who measures it. If the reader is up to the task, consider light from a distant star traveling towards two different spaceships, each traveling at a different speed towards that distant star. Imagine two photons (a discrete particle or packet) of light that have been traveling side by side. In this "thought experiment," the quicker vessel has just caught up to the slower one. At that moment, each photon encounters a different vessel. If both ships are traveling in the same direction with one traveling faster than the other, how can they both measure the distant star's light at 299,792,458 m/s? The answer is by abusing mathematics or by wearing some type of blinder.

Newton's Second Axiom

"Acceleration is directly proportional to force and inversely proportional to mass."

The relationships between force, mass, and acceleration do not change; the only substantive change to Newton's Second Law, besides the name and its wording, is the removal of references to direction from force and acceleration. These get replaced with scalars along with the understanding the word "force" refers to the concept called energy. Newton's inclusion of a directional component only happened because Descartes formulated momentum to meet an erroneous supposition.

The wording for this axiom describes Equation 30. In it, neither force nor acceleration are in bold type since they are no longer vectors. The more familiar looking, *F = ma* equation, may or may not be easier to remember but, Equation 30 is superior; it communicates the relationship between force, mass, and acceleration in a far better way. This is because it directly shows that when force increases, acceleration follows suit; it increases. It also shows that when mass increases, acceleration decreases.

$$a = \frac{F}{m}$$

Equation 30

Please note the act of applying force to an object does not necessarily mean that it will accelerate. This is because the word "force" describes energy that *attempts to or causes* acceleration. For example, a person can push on a shopping cart, and it will accelerate. However, if the cart is against a wall, the cart will remain motionless. Some things require more force than others to move for various reasons; walls generally need machinery to move them. If a shopping cart is in the middle of an aisle and has two people pushing it in opposite directions, it will only accelerate based on the net force it receives. If one fellow pushes with ten newtons and the other pushes with only nine, the cart will accelerate towards the weaker fellow based on a net force of 1 newton. In short, objects only change their motion based on unbalanced forces or the effects created by gravity.

Newton's Third Axiom

"To every action there is always an opposite reaction of equal magnitude."

This third axiom basically only requires condensing it and having a clear understanding that "action" and "reaction" are words referring to the effects energy creates. This axiom does not necessarily demand equal changes in motion or the same effects, just equal applications in the amount of energy.

Once again, take the case of an automobile slowing down as described in Scenario 2 in Chapter 3; the car slows down, but the brakes heat up. In other words, one thing experiences a significant change in motion and something else experiences an entirely different effect.

This seemingly minor change to Newton's Third Law is more significant than it might first appear. Newton's original wording foreshadowed the development of chemical rockets which, eventually led to sending man to the moon. This correction foreshadows a great deal more. For example, it does not prohibit the construction of spaceships that do not use propellant in the usual way. And without the need of ejecting propellant, a spaceship could travel far faster than chemical rockets. A trip from Earth to Mars might one day be no different than traveling from California to Europe.

A natural consequence of this axiom is that when force acts between two objects, they both experience force for the same amount of *time*.

Next item up for discussion is a brief treatise on the Law of Conservation of Energy. There the reader will come to learn that there is a bit more to understanding energy conservation beyond the simple statement that it is. Spoiler alert; think companion laws.

CHAPTER 12

The Law of Conservation of Energy

The concept of conservation holds a particular importance in physics; without it, it is not possible to unlock the nature of nature. This is because when some property does not change during an event, physicists have a point from which they can compare situations. Philosophically speaking, it is only possible to understand something if there is something else with which to compare it. For example, if someone learns that a meter is a unit of distance but has nothing with which to compare that distance, he will have difficulty understanding it. If he knows how long a foot is and learns that a meter is slightly more than 3 feet, he will find it far easier to grasp what a meter is.

Descartes popularized conservation. He supposed that the amount of motion pre- and post-collision remained unchanged. Decades later, Leibniz assumed Vis Viva also had the status of a conserved quantity. That belief eventually transformed into today's Law of Conservation of Energy. Several decades later, French scientist Antoine Lavoisier (1743 – 1794) hypothesized mass might be conserved. He then tested that idea and found his suspicions were indeed valid. More than a hundred years after that, conventional wisdom dictated we lived in a static universe, a universe that was neither expanding or contracting and thus, the conservation of space. Spatial conservation remained in vogue until American astronomer Edwin Hubble (1889 – 1953) came along. His observations led to today's understanding of an ever-expanding universe and thus the continual creation of space. More than half a century after that, momentum's conserved nature was tested by someone who did not assume it was true. As for energy, invariable rules exist regarding its conservation but, that does not necessarily mean that energy can never be created or made to vanish.

The fact that energy can be created should not surprise physicists. Consider the amount of energy in the universe; it is absolutely staggering. Just the dynamic energy of the moon as it makes its way relentlessly around the Earth is numerically overwhelming. Try multiplying its mass of roughly 73,460,000,000,000,000,000,000 kg by its average orbital velocity of 1,022 m/s. It works out to a pretty big number to say the least and that is just one moon traveling around one planet in a vast universe. For comparison, a bullet leaving a handgun typically has less than 5 bobs of dynamic energy. This is enough energy to end a human life. Interestingly, the energy applied to the handgun is the same. The bullet is far deadlier because when it strikes someone, it attacks a smaller area much faster.

The thing is that all the energy in the universe came from somewhere somehow at some time. It is beyond reason to think that all the energy and matter in the universe has been around forever. Even today's physics community understands this; they are fond of pointing to their own unique creation myth called The Big Bang. It states that some 15 billion years ago out of nothing, a great event produced all the matter and energy in the universe. This means physicists believe in two contradictory ideas; energy cannot be created out of nothing, and energy was once manufactured mysteriously in the vast nothingness of empty space.

Vanishing Energy Part1

The Cart Experiment contains within it an example of energy vanishing. That fact was not mentioned before but, it may have been noticed by some readers. This happens during Phase 2 and it is the energy related to the cart's motion that disappears. That event serves as the perfect case study; it clearly illustrates how it is possible for energy to vanish. Quite surprisingly, it also explains why this phenomenon makes perfect sense. This may seem like an impossible task and yet, it is quite simple to explain and understand.

To begin the analysis, we will state that the combined mass of the cart (m_{cart}) and its attached mechanism is 2 kg. This value is a bit unreal, but its use allows for simpler calculations that are easy to follow. Recall from Chapter 7 that the projectile's mass ($m_{projectile}$) was 2 kg. The value for the initial speed ($s_{initial}$) of the cart and projectile moving as one as Phase 2 begins is 5 m/s.

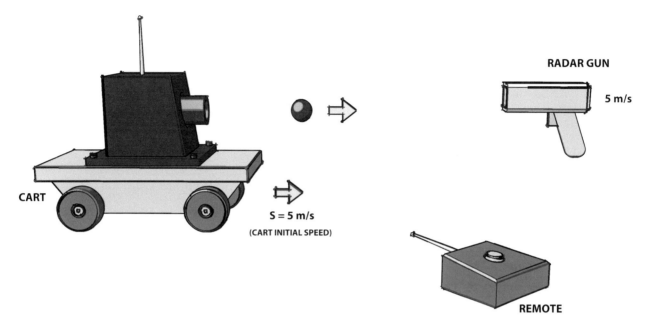

Phase 2 of Cart Experiment

Please note that in the interest of keeping things simple, we will assume that no energy "losses" exist in the following example.

$$DE_{cart\&projectile} = (m_{cart} + m_{projectile}) \times s_{initial}$$

$$DE_{cart\&projectile} = (2 + 2) \times 5 = 20 \text{ bobs}$$

Energy of Cart and Projectile

In words, the cart with its attached mechanism and the two-kilogram projectile have a combined mass of 4 kg. As Phase 2 begins, they move at 5 m/s and so, together they have 20 bobs of dynamic energy ($DE_{cart\&projectile}$).

As Phase 2 proceeds, the cart's mechanism operates and causes 10 bobs of effort to act on the projectile. When this occurs, 10 bobs of effort also act on the cart; this is Newton's Third Axiom in play.

Before using the cart's mechanism, this isolated system had a total of 20 bobs of energy. After adding 20 additional bobs (10 to the cart and 10 to the projectile), the two-kilogram projectile doubled its speed, and the cart stops moving. However, when the cart's mechanism finishes operating in Phase 2, this system still only has 20 bobs of energy, not the 40 one might think it should have. The cart is at rest and so, it has no dynamic energy. The two-kilogram projectile, now moving at 10 m/s, has 20 bobs of energy (2 kg × 10 m/s = 20 bobs). Where did the added 20 bobs of energy go? Did they simply vanish?

The answer is basically yes and for a good reason. Consider what energy is; it is *that which causes or creates effects*. It takes 10 bobs of energy to *cause* a two-kilogram cart and its attached mechanism to change its speed by 5 m/s. After 10 bobs of different energy act in the opposite direction, the cart and its mechanism return to rest. This means it no longer has the 10 bobs of energy it had before Phase 2 began. This means the energy applied to the cart during Phase 2 *undid* the effect other energy had previously created. This is like paying someone $10 to move a car from one place to another and then paying someone else $10 to move it back. All the money spent for this endeavor is gone and the vehicle is back where it began. This is not a perfect analogy but, it begins to approach perfection when observed solely from the perspective of the person doing the funding. Humorously, if the two individuals who did the deed burn the ten-dollar bills they receive, that money would also vanish from the universe.

The physicist who finds any of this unbelievable should review the mathematics and physics he was taught to explain the Cart Experiment. In Phase 1, the cart's mechanism does positive work to both the cart (having a mass of 2 kg) and the projectile; both accelerate from rest to 5 m/s but, in opposite directions. This works out, pun intended, to 25 joules being added to the projectile; the cart also receives 25 resulting in a total of 50 joules. This means the cart's mechanism must always add 50 joules of energy. This is as mandated by the existing Law of Conservation of Energy.

In Phase 2, the same mechanism used in the exact same manner now magically does 75 joules of work to the projectile. At the same time, the mechanism does 25 joules of *negative* work to the cart. This is according to Formula 15 from Chapter 5 (**Fd** $= \Delta\, \frac{1}{2}m\boldsymbol{v}^2$). This adds up to the same net amount of work as before (75 + (–25) = 50 joules). However, in Phase 1, the math looks differently; it manifests as 25 + 25 = 50. This *changing distribution of work* does not make rational sense.

Physics students do not find any of this odd. Typically, they only calculate the kinetic energy of the entire system before and after the cart's mechanism operates. To wit, from Phase 1 they know the cart's mechanism adds a total of 50 joules. When Phase 2 begins, they know that together the cart and projectile have 50 joules of energy. As the cart fires the projectile during Phase 2, the cart's mechanism adds another 50 joules to the existing 50 in this closed system. According to the Law of Conservation of Energy as physicists understand it, this means that there must be 100 joules of energy at the end of Phase 2. And there is according to the kinetic energy formula; the cart has none because it is at rest and the projectile has 100 joules (a 2-kilogram projectile moving at 10 m/s = $½mv^2$ = ½ × 2 × 10 × 10 = 100 joules).

Students believe they are being well taught because the math seems to work out so perfectly and so, they carry this understanding throughout their physics career. Unfortunately for conventional wisdom, negative work is not a valid physics principle. This would be so even if positive work passed every test it faced. Those who think otherwise, must read Appendix F, "Negative Work," before continuing; it explains why the idea of negative work is, to be blunt, ridiculous.

The question physicists might ask of themselves goes to plausibility. Are the changing distributions of work more plausible than the idea that energy creates observable effects that can be undone by other energy? Occam's Razor may very well apply in this situation; it is a philosophic problem-solving principle that goes back to a 14th century Franciscan friar. Occam's Razor basically states that the simplest explanation of two competing ones is usually the correct one. With physics as she is taught today, one must be comfortable with weird or magical transfers of energy that vary with velocity; not to mention the ridiculous notion of negative work. With the Dynamic Effort Hypothesis, you need only accept that energy can create an effect that can be undone by other energy.

Vanishing Energy Part 2

The Cart Experiment makes the idea energy can vanish somewhat palatable. Obviously, this first requires verifying mechanical energy is as described by the Dynamic Effort Hypothesis. It should become easier to accept when one accepts that energy is the concept of cause, something that creates effects or change.

With energy defined by work (force × distance), physics instructors never have any difficulty getting their students to accept their lessons. A popular one has a teacher holding a book or some other item in his outstretched arm. He goes on to state that because the book is not moving, gravity does no work. This means no energy acts on the book; it has a force acting on it but, that force does not act through a distance. If the book falls, it pushes air around and makes a noise when it collides with the ground. These effects are said to be the direct result of the work done by gravity; force accelerates the book through a distance. The explanations for why the instructor's arm eventually tires always satisfies students. In truth, the instructor's arm tires because he is continuously using biological energy against the energy gravity applies.

When the physics instructor holds the book in his outstretched arm, two opposing forces constantly act on it. One of these is due to the instructor and the other is due to gravity; consequently, the book does not fall. This is because the two opposing forces cancel one another. If either force were to vanish, the book would accelerate in the same direction as the remaining force acts. So, if the effect of gravity were to vanish suddenly and inexplicably, the book would accelerate upwards. The corollary to this is that if the instructor stops pushing the book up, it falls. And because a quantity of energy is related to the time force acts, energy constantly vanishes in this scenario. In other words, anytime two equal and opposing forces act against one another, energy vanishes. This is the same situation that occurs to the cart in Phase 2 of the Cart Experiment. In that case, the opposing forces occur at different times; in the case of the instructor holding the book, they occur simultaneously.

Please note that to be completely accurate, gravity does not produce a force of attraction; something else is going on. Gravity itself is not an illusion but, the idea that gravity produces a force of attraction is. If gravity existed as an actual force, elephants would be able to fly, and birds would be stuck on the ground. This would occur due to Newton's Second Axiom, $a = F/m$. Birds have hardly any mass compared to an elephant and so, they would accelerate towards the ground extremely quickly. An elephant would be able to leap from a cliff and gently float back down to the ground without any care or fear. Despite this, thinking gravity is a force of attraction can be helpful but, it is valid only in certain limited circumstances.

Imagine an object is on the floor and the goal is to lift it 1 meter. If this object weighs 10 newtons, the exertion of a 10-newton force acting against gravity's grip accomplishes nothing; the object remains motionless. Recall that weight is not how much mass an object has but rather the force it exerts due to gravity. Without getting overly complex, the usual action of lifting an object begins by briefly exerting a force greater than the object's weight. This is then followed by momentarily reducing the force until the object is moving at the desired upward speed. This then requires exerting the same amount of force that gravity does. In this way the object moves at a steady rate, no change in the object's speed because the opposing forces are equal. At this point, the object obeys Newton's First Axiom. As the object approaches the desired 1-meter mark, the amount of force applied in the upward direction is lessened. When the object briefly stops, the 10-newton upward force is reapplied, and the object remains stationary. When a person picks something up, he goes through these or similar changes in force quite rapidly and automatically since he has been doing this his entire life.

A fascinating fact occurs when lifting an object from one position to another. The force applied varies throughout as just described and yet, the average upward force is equal to the force gravity applies during that period. If it takes 2 seconds to lift an object 1 meter, the total energy required for that endeavor is equal to the energy it takes to hold the object for 2 seconds. A small difference in energy does however exist between lifting an object for 2 seconds and holding it in place for that time. That energy difference is due to air being pushed about during the lifting process.

If the preceding sounds strange, consider the following. In space beyond the limits of Earth's atmosphere and any gravitational effects, it takes just a tiny bit of energy to get an object moving. Once that is done, the object can move ten meters, a hundred, or ten billion without adding any more energy. The time for the object to move each distance may be far different but the energy requirement is not. On Earth, it takes a great deal more energy to drive a car ten kilometers compared to driving just one. This is because the other laws of nature get added to the mix; cars must deal with gravity, wind resistance, and so on.

Physicists will likely find this energy vanishing business quite alien. They will not want to believe it and yet, it must be true if the DEH is valid. Physicists need to remember that strange things happen all the time on this planet. The revelation of vanishing energy is just another scientific example in a long line of them.

Energy Creation

Hypothetically, a billiard table can demonstrate energy creation. Consider the following situation using the three billiard balls lined up as shown.

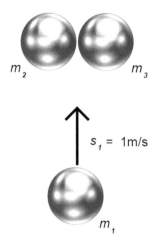

Pre-collision

Before the collision, m_1 has a certain amount of dynamic energy. A typical billiard ball has a mass of about 0.15 kg. To make the math simpler, we will use ones that are significantly larger; they have a mass of 2 kg. If one such ball was moving at 1 m/s, it would have 2 bobs of dynamic energy ($m_1 \times s_1 = 2 \times 1 = 2$).

If the balls are correctly placed and m_1 collides just so, we would see a result much like that shown in the following post-collision graphic.

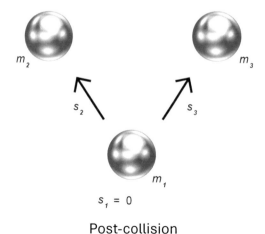

Post-collision

When the collision occurs, the original 2 bobs of energy moves from m_1 into the other billiard balls; they each acquire some dynamic energy. A far smaller quantity transforms into thermal energy; m_1, m_2 and m_3 heat up slightly. Moreover, a clicking noise also occurs which, is evidence of acoustical energy. Both the thermal and acoustical energy manifestations qualify as energy "losses;" these we will note with a single E variable that also includes the energy causing a bit of wind.

If the balls leaving the collision have the same mass as m_1, they would be traveling at about 0.7 m/s. That value is found using trigonometry; the angle in play is 45 degrees and the cosine of that angle is 0.7071068. In this situation, 0.7071068 m/s represents the maximum hypothetical speed the two balls could travel. However, if we include the "E" variable to make the situation more realistic, it becomes possible to say each of the balls leaving the collision travels at 0.7 m/s; this keeps the following math simpler.

The amount of energy after the collision occurs is calculable as $m_2 s_2 + m_3 s_3 + E$. We know that m_2 and m_3 will be moving at the same speed of 0.7 m/s and the mass for each ball is 2 kg. Using these values, we can calculate the DE_{final} after the collision.

$$DE_{final} = (m_2 \times s_2) + (m_3 \times s_3) + E$$

$$DE_{final} = (2 \times 0.7) + (2 \times 0.7) + E = 2.8 \text{ bobs} + E$$

Final Energy Calculation

The amount of dynamic energy before the collision was only 2 bobs. After the collision occurs, the amount of observable energy is more; it is 2.8 bobs plus the tiny bit represented by the E variable. This is a significant energy increase. In terms of percentage, this works out to about a 40% increase in energy.

If someone wanted to build a machine on this idea, they would need to know a bit more. For example, they would have to consider the fact that the energy required to get m_1 moving at 1 m/s is a value twice as great as noted (Newton's Third Axiom). This translates to 4 bobs (2 + 2). So, the machine would need to send two balls in opposite directions and harvest energy from two different sets of collisions. Thereafter, the machine would have to be reset and that would be where the machine's practicality ends. A single event does however produce more energy than it consumes.

Permanent magnets are also an example of energy creation. They demonstrate this whenever they are used in situations like putting a child's drawing on a refrigerator. Any non-magnetic object of similar mass falls when placed in a similar position. When magnets are placed on a refrigerator, they exert an attractive force. If the object placed against a refrigerator is not a magnet but can be turned into one using electricity, it may become a little easier to see that permanent magnets generate energy by merely existing. An electromagnet requires voltage and thus current to produce a magnetic field. They will only stick to the refrigerator during the time electrical energy is in play. From an earlier chapter, a quantity of electrical energy is equal to the applied voltage multiplied by the current and time.

Interestingly, many a physicist has debunked attempts to create free energy. This often occurs by merely citing the Laws of Thermodynamics. Given the revelations in this book, one must wonder if any physicist will do the "impossible" and devise a method of producing "free" energy. A confirmed DEH means the laws of physics no longer forbid such success. Beyond issues with existing Patent regulations, lays an unfortunate issue. Criminals and other evil doers will seek to exploit others and will falsely claim success. Charlatans have been doing this for centuries and with the publication of this book, they may be emboldened.

Understanding why energy creation is possible begins by knowing that mechanical energy is strictly a one-dimensional phenomenon existing within a three-dimensional universe. Despite this, energy can manifest in two or three dimensions of space without increasing or decreasing in quantity. For example, a spinning disk basically functions in two dimensions and would spin indefinitely if no energy were removed at its axle or otherwise. Similarly, for a planet like ours, it is a three-dimensional object that spins in the void we call space; it too seems to store energy in more than one dimension. In each of these scenarios, we have the condition of an object moving in a straight line while a different condition is in play. One condition is the bond molecules at the outer most regions have with the molecules at the center and all the molecules that exist between them. Then there is the concept of gravity for orbiting bodies such as the ISS and the moon. The combination of two or more conditions is why something can move in a circular, elliptical, or some other non-one-dimensional manner.

When a one-dimensional aspect of energy splits into two dimensions, it can result in energy creation. The corollary to this is that should the reverse happen, energy can vanish. For example, if the two billiard balls (m_2 and m_3) were to move toward and collide with m_1 at the exact same time, some quantity of energy would vanish but not all. This is because there would be an energy "loss" within the balls; remember, an energy "loss" is where energy transforms into an unwanted form such as heat.

In short, energy can be created simply because a one-dimensional phenomenon can be split into two dimensions. Moreover, energy can also vanish when using energy to undo the effect other energy previously created.

Some or most physicists may have difficulty with this energy creation and or vanishing business. This is because they believed the reverse was true for so long. When Leibniz first proposed Vis Viva, his philosophic bent did not allow him to think Vis Viva could be created or destroyed. His assumption easily found acceptance and quickly became something everybody took to be true; no one considered that it might not be so.

As physics became more organized, certain problems with work and kinetic energy came to light. This caused physicists to add absurd things like negative work because energy "cannot be created or destroyed". They also relied on momentum in certain situations because the kinetic energy formula did not function within their understanding of conservation.

Gravity, magnets, and a few other phenomena seem to continuously produce energy which, is often countered by other energy at the same time. The reason for these phenomena goes to a different but related hypothesis. Unfortunately, it has not been fully developed and even if it was, it is well beyond the scope and purpose of this book. It will not be explored or mentioned again until the physics community confirms the DEH, and probably not even then. If the physics community cannot handle the undeniable corrections to physics this book provides, they will really have a time of it trying to deal with the details of the grandest illusion of all time, bar none.

The author struggled with adding this chapter but ultimately, it had to be included. Any physicist worth his salt would sooner or later recognize that energy vanishes during Phase 2 in the Cart Experiment. The difficulty with adding Chapter 12 goes to the previous paragraph. The design of this universe requires certain laws exist. Some of these will not appear to make sense given the history of physics. Uncovering the laws of nature is a marathon and revealing too much too soon will trip up a great many. Hence, this chapter only reveals the bare minimum.

Please note: the author is not asking for a leap of faith. Test the DEH's validity and everything else; see if energy increases directly with speed or not. Thereafter, see if energy can be created and or made to vanish.

A Few Final Words

This book provides enough information to jump start a golden age in physics. It only requires that the physics community remove their blinders. Unfortunately, this is not easily done; sometimes a cultural or scientific blinder takes decades or longer to dissolve. On the other hand, those physicists who venture on ahead just might use the revelations herein to produce new technological wonders; stranger things have occurred.

APPENDICES

A. Secrets to Understanding
B. Mathematics Simplified
C. The Scientific Method
D. The Academic Experiment
E. Momentum and Kinetic Energy
F. Negative Work
G. Historical Figures

APPENDIX A

Secrets to Understanding

Public schools do not teach students how to understand; they only teach students the mechanics of how to read. There are no public-school courses that teach students the secrets to understanding. In the classroom, the teacher speaks and assumes students are learning. Some knowledge is passed on in this way but, it is far less than anyone suspects.

Many students study basic arithmetic and later algebra. After a few years go by, they have trouble with any math beyond simple arithmetic. Some adults on seeing a simple equation with a couple of variables immediately decide the material is too advanced and forgo reading it. In general, adults remember and can use only a small percentage of the math and other subjects they studied. This is regardless of how well they did on their exams or the grades they received. This is due entirely to the fact students are not taught any of the secrets to understanding. As a result, they do not retain most of the information they read or hear. To ensure anyone can comprehend the author's research, this appendix teaches how to understand anything.

The First Step

The first thing everyone must realize is that it is possible to understand *anything*, and that includes physics. When someone thinks a subject is too difficult, it means he or she is wearing a blinder. The thing about blinders is that they prevent individuals from doing the first thing they need to do; just observe what is there. Anyone can understand anything if they are willing to learn *and* willing to observe. Learning requires looking at the information as it exists; not as something horrific to be avoided. Once that is done, understanding becomes possible using the most important study secret.

For those who already believe they know all about some subject, they must find a way to shed this particular blinder long enough to observe. When you think you know all there is to know about something, there is no point in looking and so, you do not see. Everyone needs to understand that there is always something new to learn. It may or may not be tremendously important but, there is always something there. In short, before one can learn or judge anything, it first requires the willingness and action of just looking.

The Most Important Study Secret

Because most knowledge is passed on via the written and or spoken word, it is imperative that one understands all the words being used. A single word whose definition is unknown, incomplete, or the wrong one for the context in which it is used can alter the meaning of a sentence. This can then give the reader the wrong idea and cause confusion. This can be particularly catastrophic in technical and scientific subjects. The corollary (something that naturally follows) to this is that if you know the meaning of every word in a sentence, you will understand what the author intended.

Consider the word "theory" which, is particularly relevant to physics. It has two definitions that conflict in an odd way. This can sometimes cause confusion even amongst physicists. One definition is *the principles considered a fact that describes a phenomenon of nature*. For example, the theory behind multiplication is that it is a shortcut to adding numbers. The same answer occurs when someone multiplies 3 by 4 ($3 \times 4 = 12$) or adds the 3 four times ($3 + 3 + 3 + 3 = 12$). As the numbers get larger, the value of multiplying becomes undeniable; for example, just try adding the number 33 to itself 44 times.

The second definition can be summed up with the word "hypothesis". It means a *proposed explanation for some phenomenon of nature*. Currently, some physics "theories" are unproven hypotheses. Sadly, some of them are thought to be true by some physicists and a great many laymen. And so, it may be time to retire the hypothesis definition for the word "theory."

Most English words have multiple definitions and the one that applies is the one that fits within the context of the sentence in which it is used. Occasionally a writer will use the same word in a sentence more than once where each use of the word has a different meaning; sometimes they have the same one. In other words, stay alert when studying or reading. For example, when enough scientists prove a *theory* is in fact true, that *theory* can rise in status to become a teachable *theory*. In this example, the first two uses of the word "theory" relate to the hypothesis definition. The last use of the word "theory" relates to the other definition.

The thing about words that may be misunderstood is that it can be any word, or even a symbol. It might be a physics term that a reader might not understand. However, it can also be some other word or symbol, especially if the book's author was diligent about defining the physics terms. Symbols include such things as periods, commas, and the various symbols used in mathematics.

As it so happens, whenever someone goes past a misunderstood word or symbol, clues begin appearing. The more words and symbols that someone fails to understand, the more numerous and forceful these clues become. One of these is the inability to remember what was just read. Others include failing to see certain words or seeing ones different from those on the page. Still others include thinking the text is saying something different than what it is saying, feeling dull, or having one's mind wander. There are other similar such clues but suffice it to say, do not go past those words or symbols for which you do not know the correct definition. If you go by enough misunderstood words or symbols, it can cause you to stop reading a book entirely. The corollary to this is that if you know all the definitions for all the words and symbols in a book, you will have absolutely no difficulty understanding it. You will also be able to pass any test on it and be able to use the information as if it were of your own creation.

Whenever the material being studied or read does not seem to make sense or some other clue becomes known to you, stop at that point. Then go back to when everything made sense and look for a word or symbol for which you do not have the correct definition. The one thing to remember is that the misunderstood word or symbol is often *somewhere towards the end* of the material that made sense to you. After finding the word, check the glossary or a good dictionary and find the definition that applies. Once you understand the word or symbol, resume reading the material starting with the first sentence of the paragraph in which the misunderstood word was found. Do this well and things should once again make sense. If they do not, look for another word or symbol that might be misunderstood.

As a quick aside, when trying to understand the definition of a word, make sure you understand all the words that define it. The same rules that apply to reading also apply when trying to understand definitions. And for the record, it is best to use dictionaries that do not employ seldom used or strange words in their definitions. Nothing is more frustrating than looking up a word in the dictionary only to find it necessary to look up a number of other words.

Do not try to figure out what a word means, get a dictionary, check the glossary, or get on the internet and find out. Teachers sometimes encourage students to guess what a word means because they themselves never learned how to understand; no one told them to look up words in a dictionary. On rare occasions, a teacher might mention the use of a dictionary or even give lessons on their use. They never tell students they should be using dictionaries *whenever* they read or study. The most important thing to know about learning is to know what the words and symbols mean in the text you are reading.

There is no shame in looking up any word or symbol; it is however quite shameful not to do so. Students who routinely go by words for which they do not have the proper definition tend to lose IQ points over time. A silly notion exists in our society that a smart person is one that has one or more advanced university degrees. Having studied a lot of facts does not make someone smart. A computer can store a great deal of information but that does not make it intelligent. Intelligence is about the ability to observe reality as it exists and use those observations constructively. Ironically, individuals with multiple degrees and doctorates often do not know the definitions for some rather common words such as "of," "is," and "as." However, they tend to have good memories and so, they have little to no difficulty passing exams.

You can always tell when you understand all the words and symbols when you read; everything makes sense. You can think with the information and see how it might be used or exploited. As an additional means test, you can also ask questions of yourself as you finish each paragraph or section. Ask if you remember the gist of what you just read or if you understand the point the author was making. You can even ask if you know how you might use or exploit the information.

Trying to memorize information is something readers should avoid; it can be a liability since it implies understanding. Memorization does not guarantee the ability to use information. Understanding requires being willing to see information as it exists and knowing all the words and symbols used to describe it. Proving that you understand requires being able to think with the information. For example, a person who cannot manipulate the equations in this book does not understand basic algebra even if they passed an advanced course on the subject.

Readers also need to know that typos sometimes happen and that they can make reading a bit of a chore. When they turn up, be sure you understand all the words so that you know what happened and where the typo is. Sometimes, it is not a typo; the author mis-used a word. This does not happen very often in published works; it does however occur surprisingly frequently when people talk, especially during unprepared lectures, speeches, and interviews.

Whenever you read something that conflicts with what you were taught previously, be very vigilant. Yes, the problem might be with the new information but, there might also be an issue with the information you learned previously. When conflicts of this nature arise, do two things. The first is make certain you understand all the words and the intent of the newer material. Make sure you have an issue with what was actually written, not with something else. If that does not resolve the conflict, work out how the new information might be valid and why it might not be. It is a very wise person who recognizes that man's understanding changes over time; what was considered true yesterday is not necessarily true today or, will be tomorrow.

To help the reader with words and symbols, this book has a glossary with two parts. The first part relays the definitions for the symbols used in punctuation and mathematics. The second part list the definitions for many but not all the words used in the text.

A Few Valuable and Useful Tips

Many books, this one included, begin with simple and easy to understand ideas. As the book proceeds, it introduces more and more complex ideas. This is a gradient approach to teaching or learning that suggests an important thing to know about studying. To wit, if the reader notices that he or she is getting confused or bogged down, they should stop and go back to an earlier point in the book. Think of learning or reading a book as climbing a set of stairs; if each step is done well, the ones that follow are no big deal. When the going gets tough, it should signal the reader that he or she failed to understand something earlier in the book. The remedy is not to dwell on the area of the confusion or difficulty; that never works. The solution is to look to the information that preceded the confusion. So, to make a confusion vanish, go back to where things made sense and find the words or information you did not fully understand, or missed seeing. Clear those up and move on from there. By the time you return to the area that caused a confusion, it will be as if the confusion never existed.

Please note; writers sometimes introduce an idea and then explain it in the next sentence or paragraph. Readers should therefore briefly glance ahead when the writer introduces a new concept that causes puzzlement. Sometimes the answer to a small confusion is a few sentences in front of the reader. Should this not be the case, go earlier in the text and find where the confusion began.

Graphic representations, aka photographs and drawings, can be of enormous value when reading. They make the information given in the text more relatable and thus help readers understand the author's intent. Sometimes, it is not possible to grasp something unless one has the item being studied available for inspection. For example, consider what it would be like for someone trying to understand the inner workings of a combination lock. Without the lock, or a detailed drawing or a picture of the internal mechanism, a student could easily get dazed and confused. In other words, effective learning should include at least some graphic representations or, better still, the item being read about; it makes for a better communication line between the author and the reader. If the writer does not provide anything, it is not a bad idea for the reader to find or create their own illustrations. To quote an old idea that still has merit, "a picture is worth a thousand words."

If the reader stops for a moment and evaluates the information in this appendix, he or she should be able to see it makes perfect sense. You cannot learn if you think you know all about something and so do not look. Learning requires the willingness to observe what is actually there. To understand the written or spoken word, one must understand the meaning of the words and symbols used in communicating ideas. Graphic representations, and better still, the objects associated with the text can be enormously valuable; they should be part of the learning experience. And finally, the more advanced aspects of any subject rest on the basic principles that precede it. If you understand these concepts, you can work out a blueprint to learning anything. It should always begin with a willingness to learn, start with the simple books first, know all its words, and gradually move on to more complex ones.

APPENDIX B

Mathematics Simplified

In this book, the mathematics (aka math) is quite simple and will be easy to grasp. This becomes a fact once readers do three things. The first requires readers remove any mathematical blinders they may be wearing; just get the idea the math might not be that difficult. The second is that they should use the proven techniques given in Appendix A and finally, read this appendix.

For many, math can appear intimidating because it uses unfamiliar symbols and has mysterious rules. However, mathematics becomes far less troublesome when those symbols and the rules behind them are known and well understood. Take the word "mathematics"; it is just a word that describes a subject that deals with numbers and the various ways numbers can be manipulated.

Numbers are merely symbolic representations. Some people show up for a party and when we count how many there are, we find there are 1, 2, 3, 4, 5 guests. If each of those five guests brings two bottles of wine, we can use mathematics to tell us how many bottles we have. Without the ability to multiply 5×2, we have to count the bottles to know how many. Mathematics is not an enemy; it is a friend and one that becomes a better one the more you get to know him. And the key to getting to know him is understanding what the symbols mean and the rules that apply to them. And as always, a lot of practice with those rules is a good idea.

Almost everyone can do simple arithmetic. This is just the adding, subtracting, multiplying and dividing of numbers. Today with calculators everywhere (check your smart phone), no one needs to use a paper and pen to add, subtract, and so forth; you really don't even need to know the multiplication tables or how to add. A half century ago, students and engineers did not use calculators; they did not exist to the degree they do today. That is when doing mathematics was far more challenging so, consider yourself a bit luckier.

Sometimes, the mere mention of the words "math" or "algebra" is enough to make someone's eyes glaze over. This is a testament to how poorly mathematics has been taught (see Appendix A for why). Algebra is not a horribly difficult subject; it is a very broad and general type of mathematics that uses letters to represent various things. These letters are called *variables* because they can represent any number; they are basically symbols of known and unknown numerical values.

When you see something like *a* + *b* = 10, you see an *equation*, something with an equals sign. This example has two variables (*a* and *b*). All this equation tells us is that there are two unknown numbers that when added together equal a value of 10. And when we say 10, we are really talking about 10 things be they oranges or bottles of wine. If the numbers never represent something whether stated or implied, mathematics begins to descend into a purely academic activity that has little practical value.

In the equation *a* + *b* = 10, the numbers listed on the left-hand side of the equals sign (a + b) have the same value as the number on the other side (10). In ancient times, commerce was sometimes done using a balance scale akin to the one shown here. In the mathematics used in this book, the scale must always be balanced; they have to be equal to one another. So, to keep a perfect balance, a + b must have the same value as 10.

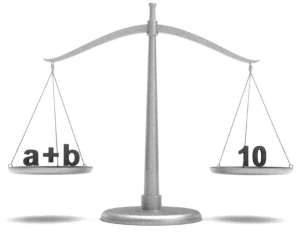

Balance Scale

Using the equation from the previous paragraph as an example, we run into the most difficult mathematics in this book; manipulating equations to make them more useful. However, when it is necessary to manipulate an equation, the author does that for the reader. If we happened to know that $b = 7$, we can figure out the value of a by using the rules of algebra to rearrange the equation $a + b = 10$. In this case, if we subtract b from both sides, we are rearranging the equation while keeping the scale balanced. This would first look like $a + b - b = 10 - b$. And knowing that subtracting a number from itself is zero, we get $a = 10 - b$. And by substituting 7 for b as indicated earlier, we now have $a = 10 - 7$ from which, using simple arithmetic, we wind up with the answer that a is 3.

This extremely simple example basically answers the question of what algebra is and how it can be valuable. In physics, equations deal with certain ideas about nature; they define relationships that can be represented mathematically using variables. Algebra gives us the means to rearrange equations thereby extending a known mathematical relationship into more useful forms.

Beyond the fact that equations exist and have value, readers should know a few other things. In this book, variables are always italicized letters of the alphabet. There are only so many letters and sometimes we need to use them for things other than variables. For example, if you see "m/s", that means meters per second, a distance divided by time indicating how fast something moves. (A meter is a distance of slightly more than 3 feet.) Neither the "m" nor the "s" is in italics. The letters "m" and "s" are also used as variables and when they are, they are written in italics which, are letters that slant to the right. In this book, the author defines the variables in the text and their meaning does not change; think about a comforting consistency.

Sometimes two variables are placed side by side as in ms; sometimes a variable is two letters long. Although this may sound like a problem, it will not be in this book. Whenever a variable is placed side by side, it will be explained somewhere in the text to mean that "m" is multiplied by "s". This can also be stated as the "product of m and s" which is the same as $m \times s$ or ms. Whenever a variable is itself two letters long, those letters will be defined as a single variable, be unique, and be easy enough to understand. In short, the use of variables in this book will not be an issue that will cause the reader any grief; the reader does not need to assume or guess about anything.

In some situations, it is necessary to use a variable more than once. For example, if a situation contains an object, the usual thing is to label it using the variable "m". It typically represents the word mass where mass is just how much stuff an object contains. In those situations, involving two or more objects, the standard action is to use the same variable for each with a different identifying subscript.

A subscript can be a number, a letter, an abbreviation, or a brief description. Subscripts are always just to the right and slightly below the variable to which they belong. For example, the reader might see a situation containing two objects labelled m_1 and m_2. And if they are moving, m_1 could use the variable s_1 while m_2 would then use s_2. In such situations, the variable "s" represents speed, how fast something moves.

Earlier, the equation "*a + b = 10*" was rearranged using one of the usual mathematical operators (+, –, ×, ÷). In that example, the variable "*b*" was subtracted from both sides. Multiplication (×) and division (÷) are the operators whenever equations are rearranged in this book. This is due to the nature of those particular equations.

The usual symbol for division in this book is not the divided by sign (÷) but, it is used occasionally within the text. Division will most often be shown in the form of a fraction. A fraction is simply one number above another as in which is the same as dividing *a* by *b* or *a* ÷ *b* and occasionally represented by a/b. In no case is it acceptable to divide by zero; to do that is like asking how many *nothings* are there in a something. There is no answer; it is undefined. Interestingly, any value or number multiplied by zero is zero.

The sequence of manipulating c × d = 20 into a different form is best understood by observing the following. The idea in this example is that we want to know the value of "c".

$$c \times d = 20$$

Original Equation

This equation states that when two unknown numbers (*c* and *d*) get multiplied, the answer is 20.

$$\frac{c \times d}{d} = \frac{20}{d}$$

Dividing both sides by *d*.

$$c \times \frac{d}{d} = \frac{20}{d}$$

Produces the following.

$$c = \frac{20}{d}$$

Final Equation

Because *d* divided by *d* is equal to 1, we get the preceding final equation. This is because any number multiplied by "1" results in no change of that number; for example, $c \times 1 = c$. In words, the final equation says that the value of *c* is equal to the number 20 divided by the value of *d*. If *d* were to have a value of 5, the value of *c* would be 4.

In this book, the various steps used to manipulate or rearrange equations will not be shown. The poor physicist already has to wallow through numerous explanations for the layman and so, we give him a break.

In this book, squares and square roots are of the simplest type. Using the example y^2, the variable *y* has the number 2 above and to its right. That "2" is the only number that will be used in that position in this book. It means that the variable *y* gets multiplied by itself. Stated in a different way, y^2 is the same thing as $y \times y$. As for square roots, they will only take the form of $\sqrt{z}$ where *z* is any number or mathematical expression greater than zero. Take the case of $\sqrt{9}$; the answer is 3 because $3 \times 3 = 9$. Shown differently, $\sqrt{3 \times 3}$. In simplest terms, a square root is the opposite of a square. When numbers are squared (y^2), the result is a value equal to the value of the number times itself ($y \times y$). In the case of a square root, the result is a value that when multiplied by itself produces the number under the square root sign.

Certain quantities or variables in this book are composed of two different parts or components. These quantities are called *vectors* and consist of an *amount* (magnitude) and a *direction*. Velocity is one such quantity and the usual variable for it is "***v***". This variable is not only written in italics, it is also in bold type. The bolding of a variable in this book tells you that the variable is a vector; it has two components. One is a quantity, aka *a magnitude*, and the other is some notation of direction. If someone is driving north at 60 miles per hour (mph), we know their velocity. If their direction is unknown, we can only say that his *speed* is 60 mph. Speed is velocity without any notation or mention of direction, just the magnitude. In this book, the letter "s" represents speed and when used in that way, the variable will be in italics but *not* in bold type. Pure quantities, like speed, time, or mass are scalar values; they have no direction associated with them. Scales tell us how much of something exists but they say nothing of direction; scale translates to scalar.

The noting of direction can take several different forms. This book mostly uses the simplest one and it consists of a positive and negative direction. Typically, anything that is shown going to the left is negative and anything to the right is positive; this is an arbitrary designation that could go the other way without anything of importance changing. There are also other more complex mathematical methods of showing direction. Fortunately for readers, these are used quite sparingly and not in any way that will confuse the reader.

Direction wise, the use of the term "negative" is not supposed to indicate that something has a value of less than zero. If the velocity of an object is given as negative 2 meters per second (− 2 m/s), an object moving at a velocity of positive 2 meters per second (+ 2 m/s) is moving just as fast; they are just moving in opposite directions. If we were tasked with adding these velocities together, the answer would be 0 m/s (− 2 + 2 = 0).

Several rules exist regarding how vectors are manipulated mathematically and they are described in the simplest terms. The first is that when two vectors are pointing in the same direction and are multiplied by one another, the mathematical result is not a vector; it is a pure quantity. For example, if some physics formula used velocity and it had to be multiplied by itself, the resulting mathematical value would not contain any directional notations. If you see $\boldsymbol{v}^2$ or $\boldsymbol{v} \times \boldsymbol{v}$, the answer would be a pure quantity (a scalar), no direction in the result. The variable $\boldsymbol{v}$ represents velocity; how fast something moves and includes the direction it travels. The same thing occurs when dividing one vector by another. So, the result of $\boldsymbol{v} \div \boldsymbol{v}$ is a scalar; a pure quantity with no hint or mention of direction.

The second rule readers need to know is that when a vector is multiplied by a pure quantity (a scalar value), the result is a vector pointing in the same direction having a different numerical value. For example, if the velocity of a vehicle was negative 60 mph and we had to multiply that by 2, the answer would be negative 120 mph. If you see something like $n \times \boldsymbol{v}$ or $n\boldsymbol{v}$, the answer would be a vector (the variable "n" is a scalar quantity and the variable "$\boldsymbol{v}$" is a vector).

The third rule readers should be aware of is that in any equation involving vectors, a balance must exist. If there is a single vector on the left-hand side of the equation, there must be a single vector on the right. In manipulating such an equation, it can occur that there are no vectors on one side and a couple on the other. In such cases, one of two situations exist. From the first rule above, we know that when two vectors point in the same direction and are multiplied together, the result is a pure quantity with no direction attached. If one of the vectors is divided by the other, the result is again a pure quantity. If direction vanishes, the equation remains balanced with regards to vectors; both sides of the equation are scalar.

Earlier the word "equation" was introduced and in this book another word is occasionally used that describes a similar concept. That word is "formula" and it is defined as an equation whose purpose is to calculate an unknown quantity.

Occasionally, an equation or formula contains parentheses. For example, if the reader sees something like 4 × (2 + 3); the proper action is to do the addition within the parentheses first and then multiply the result by 4. In this example, the answer is 20 because 2 + 3 = 5 and when that value gets multiplied by 4, the answer is 20. A different answer results if the parentheses were not there. In that case, the first action would be to multiply 2 by 4 and then add 3. That value would be 2 × 4 = 8 to which the addition of 3 results in a value of 11. The use of parentheses is to make sure anyone doing the calculations ends up with the correct result.

The reader should also know the word "factor"; it has two useful definitions. Mathematically, it describes something that is multiplied by something else. For example, if we *factor* in time, it means we multiply something by time. If we need to factor in the variable g for some reason into the equation $c \times d = 20$, we wind up with $c \times d \times g = 20 \times g$. Remember, to keep the equation balanced, we must do the same thing to both sides of the original equation. The non-mathematical definition for the word "factor" refers to something that has an effect or is an important part of something. For example, a player's size is an important *factor* for certain positions in football.

There is of course quite a bit more to know about mathematics but, if the reader understands everything in this appendix, they should have no difficulty with this book. In the interest of making sure everyone can be a witness to certain historic revelations, many of the lessons given here in basic mathematics are repeated throughout the book. And even if the reader has some misgivings about their mathematical abilities, they are not likely to have any difficulties with this book; just be mindful of the explanations given along the way while paying close attention to the variables in play. If you do nothing more than note which variables are being used, that information should be enough to grasp the author's intention.

APPENDIX C

The Scientific Method

The *scientific method* can be presented in a number of different ways. Some versions may have ten or more steps requiring a flow chart to understand; others are far simpler. The author prefers reducing the scientific method down into just four essential steps plus one important caveat.

The Scientific Method: Recognition, Formulation, Prediction, Testing

1. *Recognition*; seeing a phenomenon with no known theory to explain it or seeing something in an existing theory that seems odd and may be worth exploring.

    The Scientific Method requires a starting point and what can be more scientific than noticing something odd or unknown. Some versions begin by asking a question which, aligns quite nicely with one of this book's two themes (questions and blinders).

2. *Formulation*; gathering enough information to come up with a hypothesis to explain the phenomenon.

    Before coming up with a proposed explanation describing a phenomenon of nature, it is first necessary to have enough data. No one can offer a plausible explanation if they do not have sufficient information. Once a scientist has enough information, formulating a logical hypothesis becomes possible.

3. *Prediction*; working out a testable prediction that if confirmed provides evidence for or disproves the hypothesis.

    This is the *first* test a hypothesis must pass; does it provide some sort of a verifiable prediction? The prediction could be the description of an unknown phenomenon or it might provide a quantitative value for something that can be measured. If a hypothesis makes no testable predictions or predictions that cannot be verified using existing technology, its validity descends into a matter of faith, not science.

4. *Testing*; coming up with and doing an experiment to see if the prediction is true. With an experimentally verified or observed prediction that is in fact true, a hypothesis can begin moving from a proposed explanation to a teachable scientific theory. The prediction should be unique enough to exclude any other hypotheses or explanations.

    If a prediction is found not true, the thing to do then is to go back to step 2. This requires doing enough further research to improve the original hypothesis. After that, it is back to steps 3 and 4.

The Caveat

As with a great many things, the Scientific Method is subject to human frailties. For example, a scientist may have a strong bias for his hypothesis and that fact could impair his judgment (a personal blinder). And then there is the university or group of physicists working as one that tests a hypothesis; they may have strong opinions for or against it (an institutional blinder). And so, part of the Scientific Method is having numerous other scientists do their own experiments. Supporting this caveat is the ever-present possibility that a solitary unbiased scientist, or a group of scientists working as one, could make an experimental error of one kind or another and thus saddle the science of physics with an invalid theory. With numerous scientists doing multiple experiments, the likelihood of unseen errors diminishes greatly. For the record, Du Châtelet's research (mentioned in Chapters 5 and 6) amounts to a case study demonstrating the necessity for this caveat.

APPENDIX D

The Academic Experiment

Many physicists will point to an experiment they believe confirms the formulas for work and kinetic energy. This experiment is a staple in some introductory physics courses and it is extremely impressive. In fact, the experimental results are so impressive that no one ever stops to analyze the experiment's basic premise. There is no arguing with the mathematical results and yet, the experiment's premise contains a hidden flaw so insidious that uncovering it requires a decent understanding of Illusion Theory (Chapter 10).

The following graphic shows one version of the experiment. Other versions exist and they depend on the same basic premise given below. These usually take the form of analyzing a roller-coaster. The one shown in this appendix is less entertaining but, it is far easier to comprehend.

Because the object m_2 can fall, it produces a force that pulls the object m_1 via the pulley. Object m_1 is on an air track that allows it to move horizontally with minimal resistance. And by the use of various sensors strategically placed (not shown here), this setup allows students to test the relationship of work with the kinetic energy formula.

Typical Experimental Setup

This is a clever experiment; the object m_2 begins near the pulley and has *gravitational potential energy* (this is due to the object's position and the fact that it can fall due to the effects of gravity). When it falls, it causes a calculable force to act on m_1 via the pulley. Before this occurs, m_1 is at rest and has no kinetic energy. As m_2 falls it causes m_1 to accelerate thereby increasing the amount of kinetic energy m_1 has. The greater the mass that m_1 has, the slower it accelerates. In short, this experiment demonstrates the relationship between work and kinetic energy.

A somewhat similar experiment can be done without the table, air track, and pulley. It is described below as the "Generic Version" and represents what happens when using a single object. Think of dropping an object and measuring the necessary variables using advanced remote sensors. In essence, the Academic Experiment and the Generic Version are basically the same; both test the relationship between work and kinetic energy. The Generic Version does this directly while the Academic Experiment uses a table, pulleys and other equipment. Like many technical things, the actual means to accomplish something can sometimes appear more complicated than the basic theory behind it. In this case, the seven steps of the Generic Version represent the theory behind the Academic Experiment, or the roller-coaster versions many physics instructors use.

Generic Version

1. Drop an object of known mass (*m*) from a known height (*d*) and measure its velocity (*v*) as it reaches the ground.
2. The force (*F*) acting on the object is found using the formula, ***F*** = *m**a*** = *mg*. (The variable ***g*** is the acceleration that occurs due to gravity and is roughly equal to 9.8 m/s².)

3. The magnitude of the displacement (distance) is equal to the known height referenced in Step #1.
4. Multiply the amount of force determined above with the displacement. In other words, determine the work done to the object (work = force × displacement).
5. Take the velocity of the object as it reaches the ground and its mass and plug those two values into the kinetic energy formula ($ke = \frac{1}{2}mv^2$)
6. Compare the work done from Step 4 to the kinetic energy calculation in Step 5.

When any version of this experiment is done, the results are extremely impressive. Unless a gross error occurred, the numerical values calculated for work and kinetic energy always match within the expected margin of error.

In short, this experiment appears to make an indisputable case for work and kinetic energy. Unfortunately, its basic premise contains a fatal flaw. While the experiment does in fact verify the mathematics of how work relates to kinetic energy, it does not conclusively prove that mechanical energy (the energy associated with moving objects) must be quantified by the kinetic energy formula ($\frac{1}{2}mv^2$). This becomes immediately apparent when doing an eerily similar experiment called the "Alternate Version".

Alternate Version

A. This is identical to Step 1 of the Generic Version. Drop an object of known mass (m) from a known height and measure its velocity (v) as it reaches the ground.
B. Measure the time (t) it takes for the object to reach the ground.
C. This is the same as Step 2 of the Generic Version. The force (F) acting on the object is equal to $F = ma = mg$.
D. Multiply the force determined in the previous step with the time the object falls to determine the *impulse* (Ft) applied to the object.
E. Plug the velocity of the object as it reaches the ground and the object's mass into the formula for momentum ($p = mv$).
F. Compare the values of impulse (Ft) and momentum (mv).

When performing this experiment, the results are just as impressive as before. Unless a gross error occurred, the values calculated for impulse and momentum always match within the expected margin of error.

This version also appears to make an indisputable case but this time for a different equation. This version of the Academic Experiment is *never* done to prove the impulse momentum equation is scientifically valid. And it is never used to show that **Ft** and **mv** should represent mechanical energy.

The Explosion Scenario (described in Chapter 5) illustrates why momentum cannot represent mechanical energy. This is because momentum contains a directional component. Eliminate direction from momentum and it produces a non-directional version that does pass the Explosion Scenario. This, in essence, means that the Academic Experiment, or any other version, validates the mathematics of at least three different but related equations. It does not however show which of them should represent the energy of motion. Stated differently, the Academic Experiment is functionally useless; it only validates the math of certain equations, not the physics of mechanical energy.

$$\boldsymbol{Fd} = \frac{\boldsymbol{mv}^2}{2}$$

$$\boldsymbol{Ft} = \boldsymbol{mv}$$

$$\boldsymbol{Ft} = \boldsymbol{ms}$$

Academic Experiment Equations

The Bottom Line

The Academic Experiment cannot be cited as proof for work or the kinetic energy formula. The scientific method requires an experiment that does more than mathematically link one idea (work) with another (the kinetic energy formula). It requires an experiment that also excludes other contenders as occurs in Chapters 6, 7, and 8 of this book.

APPENDIX E

Momentum and Kinetic Energy

This appendix shows how the mathematics of momentum and kinetic energy relate to one another to create an apparent paradox.

Imagine if you will two identical objects (*a* and *b*), each having a mass (*m*) of 2 kilograms. Each of these will be moving at different velocities in the following four situations. In each case, the first two-kilogram object moves at v_a and the second one travels at v_b. The sum of v_a and v_b will always be 10 m/s. Because they have the same mass, the total momentum (p_{total}) in each scenario is always the same (20 kg m/s). This consistency of momentum is due to having a total of 4 kg (2 + 2 = 4 kg) moving at an average velocity of 5 m/s. Despite this, the total amount of kinetic energy (ke_{total}) changes in each situation.

Totals are represented by $p_{total} = p_a + p_b$ and $ke_{total} = ke_a + ke_b$

Situation 1

v_a = 10 m/s
$p_a = 2 \times 10 = 20$
$ke_a = \frac{2 \times 10^2}{2} = 100$ joules

v_b = 0 m/s
$p_b = 2 \times 0 = 0$
$ke_b = \frac{2 \times 0^2}{2} = 0$ joules

$v_a + v_b$ = 10 m/s
$p_{total} = 20 + 0 = 20$
ke_{total} = *100+ 0* = 100 joules

Situation 2

v_a = 8 m/s v_b = 2 m/s $v_a + v_b$ = 10 m/s
$p_a = 2 \times 8 = 16$ $p_b = 2 \times 2 = 4$ $p_{total} = 16 + 4 = 20$
$ke_a = \frac{2 \times 8^2}{2} = 64$ joules $ke_b = \frac{2 \times 2^2}{2} = 4$ joules ke_{total} = 64 + 4 = 68 joules

Situation 3

v_a = 6 m/s v_b = 4 m/s $v_a + v_b$ = 10 m/s
$p_a = 2 \times 6 = 12$ $p_b = 2 \times 4 = 8$ $p_{total} = 12 + 8 = 20$
$ke_a = \frac{2 \times 6^2}{2} = 36$ joules $ke_b = \frac{2 \times 4^2}{2} = 16$ joules ke_{total} = 36 + 16 = 52 joules

Situation 4

v_a = 5 m/s v_b = 5 m/s $v_a + v_b$ = 10 m/s
$p_a = 2 \times 5 = 10$ $p_b = 2 \times 10$ $p_{total} = 10 + 10 = 20$
$ke_a = \frac{2 \times 5^2}{2} = 25$ joules $ke_b = \frac{2 \times 5^2}{2} = 25$ joules ke_{total} = 25 + 25 = 50 joules

APPENDIX F

Negative Work

Physics instructors teach that work can be both positive and negative. Remember, work is the product of force and displacement (work = **F** × **d**). Positive work results in objects moving faster and negative work is said to occur when force causes an object to slow down.

Formula 15 determines if the work done is positive or negative and by how much. Mathematically, Formula 15 always subtracts an object's initial kinetic energy from its final amount and thus the possibility of negative values.

$$\boldsymbol{Fd} = \frac{m\boldsymbol{v}_{final}^2}{2} - \frac{m\boldsymbol{v}_{initial}^2}{2}$$

Formula 15

Negative work may sound reasonable to physicists but, they have not analyzed it with anything other than admiration. Apart from the mathematics of the right-hand side of Formula 15, negative work can only exist under two real world scenarios. One would be a condition in which force (**F**) is negative and displacement (**d**) is positive; the other is where force (**F**) is positive and displacement (**d**) is negative. Neither of these scenarios is even remotely possible. And in case the reader does not remember one of their basic math lessons, the only time a negative result occurs is when a positive value gets multiplied by a negative one. Two negative values multiplied together produce a positive result; the same for two positive values.

Experimentally, just try to push any object but have it move towards the push; this is a positive application of force resulting in negative displacement. Tug on the object instead and see if it moves away from the pull; it won't. This would be a force acting in the negative direction resulting in a displacement in a positive one. In short, there is no condition in which the application of force is negative and displacement is positive or vice versa; they are always in the same direction. Regardless of how force acts, an object always responds by accelerating in that same direction.

Some physicists may wish to argue that force can indeed act in one direction while an object moves in a different one. For example, they might cite the cart during Phase 2 of the Cart Experiment. The cart can appear to move to the right as force acts in the opposite direction, especially if the cart's mass is greater than the projectile. The problem is that this view is entirely superficial. When the cart begins at rest in Phase 1, force acts on the cart in the negative direction and the cart moves in that same direction. This is positive work by the physicist's own accounting. In Phase 2, an identical force acts in the same negative direction but, the cart's overall motion is to the right and thus in a positive direction. The thing is that the cart moves to the right only because it was already moving in that direction fast enough initially.

Looking at the cart's motion without any preconceptions uncovers the truth; the cart moves through two different displacements concurrently. During Phase 2, one of these is the same as occurs in Phase 1 where both force and displacement are in the same negative direction. The other displacement occurs because the cart is already moving to the right in the positive direction. Each displacement is unique and due to a different condition; one is the change in motion due to force and the other is because the cart was already moving to the right.

The Silly Little Experiment in Chapter 7 shows an example of how something can move through two displacements at the same time. If instead of handing the cell phone from the back to the front, think of the phone going the other way. This would mean the phone replicates the cart's motion in the Cart Experiment. Just focus on the cell phone and its backward motion in the car as the car moves forward. The force moving the phone is also directed backwards and so, the phone has positive work done to it. This is despite the phone's net motion, as measured along the ground, being opposite to the applied force. The thing to remember is that if the phone has no work done to it, it still moves to the right. It just moves less to the right as force acts in the negative direction.

Next consider the fact that work is not intended to be directional. The fact momentum is directional is enough to disqualify it as a candidate for energy. Therefore, if negative work were real, the use of the word "negative" would mean there is less than no work done. The thing is that there can never be less than a zero value for any physical object or condition. Just try to envision *any* real-world scenario where the time between two events is *less* than zero seconds; it is not possible. How can there be *less* than no distance between two objects or even points in the universe; it is an absurd notion. The same is true for force; how can there be *less* than no push or pull on something. A quantity of energy, whether it is thought to be defined by kinetic energy ($\frac{1}{2}m\mathbf{v}^2$) or dynamic energy ($m s$) fits into this category as well; energy either exists or it does not. And because work equates to kinetic energy ($\mathbf{Fd} = \frac{1}{2} m\mathbf{v}^2$), work cannot be negative in terms of quantity.

The bottom line is that work is a silly notion easily invalidated by the Silly Little Experiment; the idea of "negative work" is even sillier.

APPENDIX G

Historical Figures

This appendix lists a number of historical figures. Most are mentioned throughout this book but, a few are not. It can sometimes be helpful to know who contributed something to today's world and when they did so. They are listed in the order of their birth while noting their generally accepted profession and a bit more about them.

Aristotle (284 – 322 BC) Greek philosopher and scientist whose work, "Physiká", is an amalgamation of several of his manuscripts. His ideas would later be adopted by the Church of Rome.

William of Ockham (1287 – 1347) English Franciscan Friar who originated the problem-solving concept known today as Occam's Razor.

Nicolaus Copernicus (1473 – 1543) Polish mathematician and astronomer who conceived of a sun centered universe and wrote a book about it.

Martin Luther (1483 – 1546) German Professor of Theology whose actions led to the Protestant Reformation (the movement within European Christianity that led to other Churches besides the Church of Rome such as Lutherans, Baptists, and others). This eventually had the effect of reducing the Catholic Church's and Pope's influence in various areas including science.

Francis Bacon (1561 – 1626) English philosopher often credited with crafting an early version of the Scientific Method.

Galileo Galilee (1564 – 1642) Italian astronomer and scientist who popularized math and experimentation in the search for knowledge. He was arguably the first modern physicist.

Dr. William Harvey (1578 – 1657) English physician to two different English kings. He was the medical researcher who discovered how blood travels through a body and the heart's role.

René Descartes (1596 – 1650) French philosopher, mathematician, and scientist who developed momentum. Considered to be the first western philosopher.

Christiaan Huygens (1629 – 1695) Dutch physicist, mathematician, and inventor who, as Descartes' student, helped develop momentum. As an adult, Huygens befriended and mentored Leibniz.

Sir Isaac Newton (1642 – 1727) English physicist and mathematician who developed revolutionary theories on gravitation and much more. First published "The Principia" in 1687 and subsequently updated and republished it twice more.

Gottfried Wilhelm Leibniz (1646 — 1716) German philosopher and mathematician who hypothesized what would one day become known as the concept of energy. He was also Newton's chief rival.

Willem s'Gravesande (1688 — 1742) Dutch Lawyer turned scientist who developed an experiment Du Châtelet would exploit.

Andrew Motte (1696 – 1734) English writer who translated Newton's "The Principia" from Latin into English.

Emilie Du Châtelet (1706 – 1749) French aristocrat and scientist who sought to test Leibniz's hypothesis. She translated Newton's "The Principia" from Latin into French.

Antoine Lavoisier (1743 – 1794) French scientist, considered to be the father of modern chemistry. Developed the Law of Conservation of Mass. He was beheaded during the French Revolution.

Count Rumford (1753 – 1814) aka Benjamin Thompson. He was an American inventor who fought for the British during the American Revolution. He helped dispel the caloric hypothesis of heat while producing cannons. Coincidently, he married the widow of Antoine Lavoisier in 1804.

Dr. Thomas Young (1773 – 1829) English polymath whose series of public lectures circa 1805 included introducing the term "energy" to replace "Vis Viva." Those lectures also foreshadowed a change in Leibniz's hypothesis from mv^2 to $½mv^2$.

Michael Faraday (1791 – 1867) English scientist who made contributions to the study of electricity, magnetism, and electrochemistry. Faraday had no formal scientific training and yet he contributed greatly to the science of physics.

Gaspard Gustave De Coriolis (1792 – 1843) French scientist and mechanical engineer who updated the mathematics of Leibniz's hypothesis. He was the first to use the term "work" to describe a force acting through a distance.

James Prescott Joule (1818 – 1889) English brewer and physicist who equated thermal energy with mechanical work. He, along with others like Count Rumford, helped dispel the caloric hypothesis of heat. Caloric was supposed to be a fluid that flowed from hotter objects into colder ones.

James Clerk Maxwell (1831 – 1879) Scottish mathematician and physicist who published several important books on electricity and magnetism. He also published "Matter and Motion" in 1877 which, detailed the same sort of information on kinetic energy that is currently being taught.

Sir Arthur Conan Doyle (1859 – 1930) English writer who created Sherlock Holmes, the world-famous fictional detective.

Albert Einstein (1879 – 1955) German physicist who equated mass with energy and made other discoveries one of which earned him a Nobel Prize.

Alexander Fleming (1881 – 1955) Scottish doctor and scientist who accidently discovered and developed the world's first antibiotic.

Robert Goddard (1882 – 1945) First American rocket scientist. Experimented with liquid-fueled and multistage rockets. He was occasionally ridiculed by newspaper reporters.

Edwin Hubble (1889 – 1953) American astronomer who made numerous discoveries including the fact the universe contains numerous galaxies. Also found that the universe was not static but expanding.

Lafayette Ronald Hubbard (1913 - 1986) American author who delved into the human condition and developed certain valuable insights into learning, logic, and other things. These helped the author during his research phase.

GLOSSARY

The symbols included in the first section of the glossary describe those used in the text. This includes the symbols used mathematically and for punctuation. Section 2 contains the definitions for some of the words used in the text.

Section 1

Symbols, like words, often have multiple uses and definitions. In this book, only the definitions that apply are given. In those cases where a symbol has multiple definitions, be sure to use the one that fits within the context of given by the text.

Name	Symbol or example	Punctuation or Mathematics	Information
Period	.	Punctuation	Tells the reader he has come to the end of a sentence or statement.
	.	Punctuation	Used after a number or letter when listing items.
	.	Punctuation	Placed after the letters of an abbreviation.
Decimal Point	.	Mathematics	Used to separate the digits that are greater than 1 from those less than 1.
Ellipsis	...	Punctuation	Three periods used to indicate words are missing.
Bolding	**Title**	Punctuation	Used to indicate a word or words represents a title or subtitle.

	v	Mathematics	A bold letter that is also italicized is a mathematical variable that contains a directional component.
Variable	*m*	Mathematics	A variable is one or more letters that represent things like velocity that have numerical values; a temporary stand-in often represented by the thing's first letter.
Comma	,	Punctuation	Tells the reader he or she should pause briefly while reading.
	,	Punctuation	Used to separate words or groups of words.
	,	Punctuation	Sometimes used to separate added additional information.
	,	Mathematics	Used to separate large numbers into groups of three digits; not used when writing a year.
Quotation Marks	" "	Punctuation	Tells the reader the words between the marks are someone said or what was written.
	" "	Punctuation	Used to show that a word or letter is being referred to as the word itself.
	" "	Punctuation	Used to enclose the titles of books and scientific Papers as well as the names of major sections of books.
	" "	Punctuation	Used to enclose words that are used to express irony or words that have a special meaning.
Semi colon	;	Punctuation	Used to separate parts of a sentence that are different but related.
Colon	:	Punctuation	Used before listing a number of items that get separated by commas.
Parentheses	( )	Punctuation	Used to enclose additional information into a sentence or question.

	()	Punctuation	Used to list the abbreviation for something.
	()	Punctuation	Used to indicate the years someone was alive.
	()	Mathematics	Tells the reader that numbers within the parentheses need to be added, subtracted, etc. before doing other mathematical operations.
Apostrophe	'	Punctuation	Used before the letter "s" to indicate possession.
	'	Pronunciation	Used by the French as accent mark.
	'	Punctuation	Tells the reader that a letter is missing.
Question mark	?	Punctuation	Tells the reader a question is being asked.
Italics	*A*	Punctuation	Letters that slant to the right used to emphasize a word or words.
	s	Mathematics	Italicized letters represent the variables used in mathematic equations and formulas. The variable "s" represents the concept of speed.
Hyphen	-	Punctuation	Used between two or more words to create a compound word.
	-	Punctuation	Used in the middle of a word when that word begins at the end of a line and continues at the beginning of the next one.
Minus sign	–	Mathematics	Minus sign used to indicate the number after the sign needs to be subtracted from the number that appears before.
	–	Mathematics	Used to indicate a negative direction.
	–	Mathematics	Indicates a mathematical value less than zero.
Plus sign	+	Mathematics	Used to indicate addition.
	+	Mathematics	Used to show direction is positive.
Multiplication	×	Mathematics	Used between two numbers or variables indicating that one number should be multiplied by the second.

Division	÷	Mathematics	Tells the reader that the number on the left is to be divided by the number on the right. Used when the mathematics appears within the text.
Division or Fraction	$\frac{a}{b}$	Mathematics	Tells the reader that the value on the top (*a*) is to be divided by the value on the bottom (*b*).
	a/b	Mathematics	Sometimes used within the text to show division or fraction.
Exponent	r^w	Mathematics	The variable "*w*" is an exponent and in this book, it will only be the number 2. The expression "r^2" means that *r* should be multiplied by itself as in $r \times r$. If *r* were to have a value of 3, $r^2 = 3 \times 3 = 9$.
Slash	/	Mathematics	Has the meaning of *for each or for every* (per) when used for the units for speed, velocity, or acceleration. For example: meters per second = m/s.
Equals sign	=	Mathematics	Placed between two mathematical expressions to indicate those expressions have the same mathematical value.
Subscript	m_1	Mathematics	The number "1" is a subscript used to distinguish one object "m_1" from a different object designated m_2. Subscripts can take the form of a number or numbers, or a letter or letters.
Delta	Δ	Mathematics	It is the Greek letter delta and is used to mean "change in".
Theta	Θ	Mathematics	It is the Greek letter theta and is used to represent the angle between two lines.
Square root			If the value of *x* is 25, the square root of *x* is 5.
Circumflex	â	Pronunciation	A symbol placed over a letter to indicate a particular sound in French.

Section 2

Like any dictionary, this glossary lists the defined words alphabetically. Unlike dictionaries, the words defined herein only contain the definitions that apply to this book or as used by the author.

acceleration	The rate at which an object's speed or velocity changes over time.
	The speeding up or slowing down of an object.
	Uses the letter "a" as its variable.
accepted	Sometimes used with the word "theory" to indicate that the explanation for a phenomenon is thought to represent reality; a proven hypothesis.
acoustic	Of or related to sound.
aka	The abbreviation for the phrase "also known as".
akin	Like, in a similar way as.
algebra	That area of mathematics that deals with simple formulas and equations and how to rearrange them into different useful forms.
alternate	A different version.
antibiotic	A chemical substance capable of slowing or destroying bacteria.
appendix	Extra information added to a book to help a reader understand parts of the book better.
atom	The smallest bit of an element such as gold that cannot be subdivided and still retain its properties.
awry	Went in the wrong figurative or literal direction.
axiom	A self-evident statement; something that does not require proof.
balk	To be unwilling to accept an idea, to stop or hesitate unnecessarily.
bar bet	A wager usually agreed to in a pub or bar where someone makes a claim that appears impossible.
behooves	To be to someone's own benefit.

blinder	Used metaphorically to indicate something that prevents a person from observing clearly. Blinders also influence a person's reactions in a negative or less than optimum way.
boisterous	A loud noisy exchange of words and or ideas.
caveat	A warning or a caution about something.
centric	At the center of something.
challenge	To test the value or worth of someone or something.
circa	About, around, approximately.
circumvent	To avoid, get around or sidestep.
companion	A word used to describe the fact that certain laws of nature can only exist as they do because another law (a companion) also exists.
composite	A word used to describe an apparent law of the universe that is comprised of at least two even more basic laws of nature.
confidants	People with whom one can safely share secrets or private matters.
Coriolis Effect	The direction of winds due to the rotation of the Earth.
corollary	Something that naturally follows a previously proven fact or idea.
debunk	Exposing the falseness within something.
degree	A unit of temperature as in it is 95 degrees outside.
	A unit describing some portion of an angle between two intersecting lines
deleterious	Being harmful or damaging.
diabolical	Reminiscent of the devil, evil trickery.
displacement	Distance with the added notation of direction (a vector). Abbreviated using the letter "d" that is written in italics and in bold type.
distance	How far it is between two points. Abbreviated using the letter "d" written in italics.
disdain	To look upon something or someone with contempt; despise.
doctrine	A religious rule the faithful must follow.

dynamic	Having to do with motion, moving.
effort	The product of force and time using the variable EF.
egregious	Bad in a glaring or flagrant way.
elastic	That which bounces in the same way as a rubber ball.
elusive	Difficult to find, achieve, or describe.
energy	Defined by physicists as the ability to do work.
	Best described as that which is necessary for change to occur; the concept of cause. Mathematically represented by the variable "E".
energy crisis	In the 1970s, the US and other nations faced a petroleum shortage causing long lines and higher prices at gas stations.
enigma	Puzzling or difficult to understand.
enticing	Very attractive, alluring, seductive.
esoteric	Having a limited or small appeal; something that relatively few would like want, use, etc.
experiment	A controlled situation designed to test whether something is true or not.
explosion	The opposite of an inelastic collision.
	A sudden and violent release of energy.
expression	A mathematical phrase that combines numbers and variables. For example, "mx" is an expression if m and x are variables. The same is true for "$5m \div 2$" or anything similar.
factor	To multiply something.
	An important part of something.
faith	To believe in something without the necessity of understanding if or why it might be true.
foisted	Imposed or compelled to accept something.
formula	An equation designed to work out an unknown quantity.
generic	Broad or general in nature.
hypothesis	An unproven explanation for a phenomenon in the universe.

illusion	An observation or data that suggests a cause that is not the correct one.
impulse	Force (as a vector) multiplied by time given mathematically as $\boldsymbol{F} \times t$.
inelastic	Opposite of elastic, no bouncing effect; references a collision where two or more objects combine into a single one.
inexorably	In a way that is unavoidable.
intimate	Describing a close association with or a familiarity
irrefutable	Cannot be disproven or denied.
ironic	Using words to get a meaning across that is the opposite of its literal meaning. An outcome is the opposite of what was or might be expected.
italics	Letters or numbers that are slanted to the right. Numbers that are so slanted are mathematical variables. Italics are used to emphasize words.
joule	Basic unit of energy equal to a force of one newton acting through a distance of one meter.
ken	Understanding.
kinetic	Having to do with motion, moving. Abbreviated "*ke*".
lament	Cry or complain.
law	An accepted theory that describes one specific and basic aspect of nature.
layman	Someone who does not belong to a profession or is not expert in it.
magic	A form of entertainment that seeks to fool an audience by way of "miraculous" demonstrations.
mass	Roughly equivalent to weight.
	How much stuff an object contains.
	Matter's resistance to changes in acceleration. Abbreviated using the letter "*m*".
mathematics	The subject dealing with numbers and their manipulation. Abbreviated "math".
mechanical	The description of energy related to motion and its causes; kinetic energy and work are forms of mechanical energy.

meter	The standard unit of length from the International System of Units. A unit of distance slightly more than 3 feet.
migrate	To wander, move around.
molecule	Two or more atoms connected in some way. The atoms can be of the same element or not.
momentum	The amount of motion an object possesses, calculated by multiplying its mass by its velocity. Represented mathematically by the variable "*p*".
mph	Abbreviation for speed in miles per hour.
m/s	The basic scientific unit for speed or velocity measured in the distance (meters) traveled per time (second).
m/s^2	The basic unit for acceleration; it refers to the change in speed defined by meters per second per second.
naïve	Lack of experience or wisdom, new to something.
palatable	Acceptable to someone's taste or sensibilities.
pedestrian	Indicating no thought or knowledge being applied as if by a novice to some observation.
petri dish	A shallow glass dish with a removable cover used to grow bacteria or other microorganisms.
phenomenon	Something that happens; an occurrence that exists. Plural form is phenomena.
physics	The name of the science that studies motion, energy, anything related to electricity, and magnetism, and the stuff that makes up matter
polymath	An individual who knows a great deal about many subjects; an expert in many fields.
potential	Refers to the energy an object has by virtue of its position
precedent	A decision or action that serves as the guide for future decisions or actions.
premise	An assumption made at the beginning of an argument or discussion.
profound	Not superficial, having great intellectual depth.

prototypical	The first or original example of something.
recanting	Taking something back, retracting from a position.
ridiculous	Something that is absurd, should be laughed at.
scalar	A word indicating a mathematical quantity, taken from the idea of a scale indicating an amount.
scenario	A situation involving certain circumstances and or changes.
science	A subject that seeks to systematically gain knowledge about some aspect of nature.
second	The basic unit of time.
SI	Abbreviation for the International System of Units that deals with kilograms, meters, newtons, and similar units of measure.
simple	Without unnecessary complications.
solely	Only, exclusively.
sought	A form of the verb "seek" having the meaning of to look for, to discover.
space	The emptiness between two points that typically contains energy and matter. The fact things are separated from one another
speed	The description of how fast something moves without regard to its direction of travel. Abbreviated using the letter "s." (The letter "s" is not always used in this way by physicists for this purpose.)
square	To multiply a number by itself. The square of 5 is 25 (5 × 5 = 25).
square root	The number that when multiplied by itself produces a value. For example, the number 6 is the square root of 36 because (6 × 6 = 36).
stultifies	Loses focus, loses initiative, and starts dying.
tantamount	Having or being the same in value or effect
test	An experiment or an observation whose purpose is to see whether something is true or not

theory	The explanation or description of why a phenomenon exists as it does. The hypothesis definition of the word "theory" described in Appendix A is not otherwise used in this book.
thermal	Related to heat, how hot or cold something is.
time	The change experienced by everyone and everything measured using clocks and calendars. Abbreviated using the letter "t".
trebles	Increases by a factor of 3.
ubiquitous	Being everywhere at the same time.
unit	Refers to the name of something that describes a quantity. An *hour* is a unit of time, a *mile* is a unit of distance, and so on.
vector	A mathematical quantity composed of a direction and a quantity.
velocity	The description of how fast something is moving and in which direction. Usually abbreviated using the letter "v" and occasionally using the letter "u".
vibrate	To move back and forth.
vice versa	In the opposite or reverse order.
vigilant	Being very careful, watchful.
vocation	A person's job, main occupation.
wanting	Deficient in some regard or aspect.
work	Refers to the distance through which a force acts. Has the units of joules.

Printed in the United States
by Baker & Taylor Publisher Services